Structured Items for O-Level Chemistry

T. P. Borrows and O. J. Simpson

Edward Arnold

First published 1979
by Edward Arnold (Publishers) Ltd.
41 Bedford Square, London, WC1B 3DQ

British Library Cataloguing in Publication Data
Borrows, T P
   Structured items for O-Level chemistry.
   1. Chemistry — Examinations, questions, etc.
   I. Title II. Simpson, Oswald John
   540'.76   QD42

   ISBN 0-7131-0333-7

By the same Author

Basic Chemistry (with C.F. Dingle and R.A. Southcott)
The Nature of Substances — Their Structure and Behaviour
The Chemistry of Metals
Multiple Choice Chemistry
Basic Chemical Calculations
Practice Questions in Chemistry
A First Science Dictionary (with D.J. Lucas and H.I. James)

Set IBM in Press Roman by 𝓐 Tek-Art, Croydon, Surrey
Printed in Great Britain by Spottiswoode Ballantyne Ltd., Colchester and London

Foreword

The time-honoured saying 'Practice makes Perfect' is as pertinent to present-day education as it has been to education in the past. Indeed, at a time when educational thinking is strongly oriented towards the development of 'problem-solving' abilities, the saying provides a valuable hint as to how these abilities may be achieved. It is only right that during his studies, the student should be given ample opportunity to develop problem-solving skills in relation to those types of task which he is likely to encounter in examinations. With this in mind, the authors have compiled this volume of 'short answer/structured questions' which cover all essential aspects of current G.C.E. O-level courses in chemistry. A particularly welcome feature is that all questions developed and designed by the authors have been carefully pretested and, where appropriate, revised in the light of pretest results to remove ambiguities and unexpected difficulties.

It is confidently expected that the volume will prove to be both a useful teaching aid for the teacher and a valuable learning aid for the student. If this expectation is fulfilled, the efforts of the authors in compiling the volume are justly rewarded.

Richard Kempa

University of Keele
June 1978

Preface

All the examining boards now set a paper of structured questions at both O and A levels of their G.C.E. examinations, and this book arises from the need to prepare students for this type of problem. The items, which have been tested in schools, are based primarily on the O-level syllabus of the University of London schools examination board and we have endeavoured to see that all parts of this are covered. However, there are also items on topics which appear in the syllabuses of other boards, but not at present in that of London, e.g. zinc, hard water, acids on sulphides, covalency, etc. Because of this, the book is suitable for candidates taking O-level chemistry with any examining board, or the chemistry paper of a physical science examination, such as that set by Cambridge.

The arrangement of the contents has been the subject of much thought and may need a little explanation. The items have been divided amongst sixteen sections according to the knowledge required in answering them. This arrangement is, in many cases, quite arbitrary, as it would be just as logical to include, say, the 'Extraction of Aluminium' in the section on Industrial Processes – or indeed in the one on Electrolysis – as amongst the Metals. Almost all the items require knowledge of a mixture of facts and concepts, so we have included with the contents of each section a list of the numbers of other items which touch on this particular topic to a shorter extent. This list is not exhaustive, but a glance through an item will quickly show the teacher its suitability for use at any particular stage.

As far as possible, unless the logic of the item demands otherwise, each item is graded in difficulty, starting with the easier parts and gradually becoming more difficult. The lettered sub-divisions, a, b, c etc., are, however, semi-independent and pupils should be encouraged to continue with an item even when there are parts they cannot do. The items vary in style and in difficulty. We have endeavoured to cover all the abilities and activities considered by the London board when setting its own papers. The authors' experience as both examiners and moderators has helped in this. The primary ability of 'knowledge' is the most heavily weighted, but 'comprehension' and 'application' – the ability to apply knowledge gained in one situation to another sphere (see item no. 8) – are given their full share. Only a relatively few arithmetical methods are needed, but their constant repetition will help their mastery. No calculators, or even logarithmic tables, are necessary: the figures are so chosen that cancellation is simple or multiplication easy, this being the modern trend.

Finally, we must thank the chemistry staff at our respective schools for trying out the items with their classes, and Professor Richard Kempa for the very helpful comments he made on many of the items in order to improve their presentation or to remove any infelicities on our part.

T.P.B., O.J.S.

1978

Contents

viii

Section 14 Sources of Materials and Energy

Section 15 Industrial Processes

Section 16 Miscellaneous Items

Answers to Numerical Items

Items preceded by an asterisk are reproduced by kind permission of the University of London, School Examinations Department.

Useful Data

Element	Atomic Number	Relative Atomic Mass (approximately)	Symbol
Aluminium	13	27	Al
Barium	56	137	Ba
Bromine	35	80	Br
Calcium	20	40	Ca
Carbon	6	12	C
Chlorine	17	35.5	Cl
Chromium	24	52	Cr
Cobalt	27	59	Co
Copper	29	64	Cu
Fluorine	9	19	F
Hydrogen	1	1	H
Iodine	53	127	I
Iron	26	56	Fe
Lead	82	207	Pb
Lithium	3	7	Li
Magnesium	12	24	Mg
Nickel	28	58.5	Ni
Nitrogen	7	14	N
Oxygen	8	16	O
Phosphorus	15	31	P
Potassium	19	39	K
Silver	47	108	Ag
Sodium	11	23	Na
Sulphur	16	32	S
Titanium	22	48	Ti
Zinc	30	65.5	Zn

Standard temperature and pressure (s.t.p.) = $0°C$ and 1 atmosphere
Room temperature and pressure (r.t.p.) = $20°C$ and 1 atmosphere
Molar volume at s.t.p. = $22.4 \ dm^3$
Molar volume at r.t.p. = $24 \ dm^3$
Avogadro constant = 6×10^{23}
1 Faraday = $1 \ mol \ e^- \approx 96\ 000$ coulombs

Section 1
Matter: its Nature and States

1. Diffusion

In answering this question you may need the following data:

Chemical	Formula	Mass of 1 mole/g	Boiling point/°C
air	—	about 29	about −190
bromine	Br_2	160	58
carbon dioxide	CO_2	44	− 78
hydrogen	H_2	2	−253
iodine	I_2	254	183

A Three test tubes with side arms were set up with a little solid iodine at the bottom of each. Near the top of each was placed a piece of filter paper soaked in starch solution.

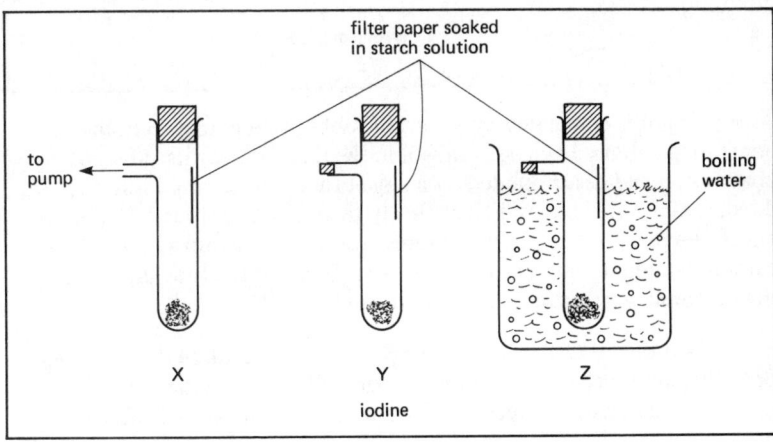

Tube X was evacuated by connecting to a pump. Tube Z was sealed with air in, and heated by standing in a beaker of boiling water. Tube Y was also sealed with air in and kept as a control. Eventually, in all three cases, the filter paper turned a deep blue colour because of a reaction between the starch and the iodine.

a Explain how the solid iodine was able to get to the starch to react with it. (1)

1

b Consider the tubes Y and Z. In which of these two would you expect the filter paper to turn blue first? Explain how you arrive at your answer.

(1)

c Consider the tubes X and Y. In which of these would you expect the filter paper to turn blue first? Explain how you arrive at your answer. (1)

d Suppose that in a similar experiment, using tube Y only, the iodine were replaced by bromine and the starch paper by a filter paper soaked in a suitable indicator of bromine. Would you expect the filter paper in this case to change colour more quickly, less quickly, or at the same rate as in the iodine experiment? Explain how you arrive at your answer. (1)

B This diagram illustrates a piece of apparatus set up for a school's Open Day exhibition.

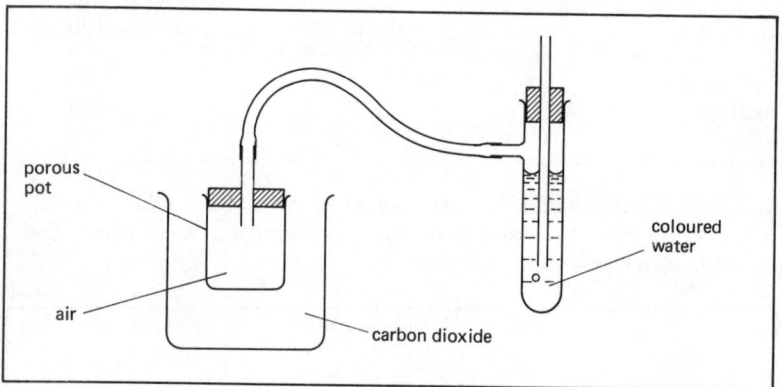

A porous pot is connected by means of rubber tubing to a test tube containing coloured water, as shown in the diagram. At first, the pot contains air and is surrounded by a large beaker containing carbon dioxide, Gases are able to diffuse slowly through the porous material of which the pot is made. A few minutes after the pot is surrounded by the carbon dioxide. bubbles can be seen coming out of the tube dipping into the coloured water, and so entering the pot.

e (i) Will the air diffuse out of the pot faster than, more slowly than, or at the same speed as the carbon dioxide diffuses into the pot?

(ii) Explain why bubbles are observed coming out of the end of the tube.

(3)

f Suppose that the experiment were repeated with air again inside the pot, but with hydrogen outside it (and the beaker sealed to prevent loss of hydrogen).

(i) Will the air diffuse out of the pot faster than, more slowly than, or at the same speed as the hydrogen diffuses into the pot?

(ii) Predict what you would expect to SEE happening to the apparatus a few moments after you had put the beaker of hydrogen around the pot. Explain why you would expect this to happen. (3)

2. Crystals

Crystals are often prepared from saturated solutions by methods which involve one (or other) of the following steps:

A a hot solution is allowed to cool
B a cold solution is allowed to stand for several days.

a (i) Explain why method **A** can result in the formation of crystals.
(ii) Explain why method **B** can result in the formation of crystals.
(iii) Which of the two methods, **A** or **B**, is most likely to result in the formation of large well-shaped crystals? Give a reason for your answer. (3)
b Give one example of crystals which may be prepared by method **A**, using a solvent other than water. State both the solvent and solute used.(2)
c State a third method, NOT involving steps **A** or **B**, which may sometimes be used to prepare crystals. Say what chemical principles are involved in the method you choose, and give one example of crystals which may be prepared by this method.   (3)
d This diagram represents the arrangement of particles in a certain crystal:

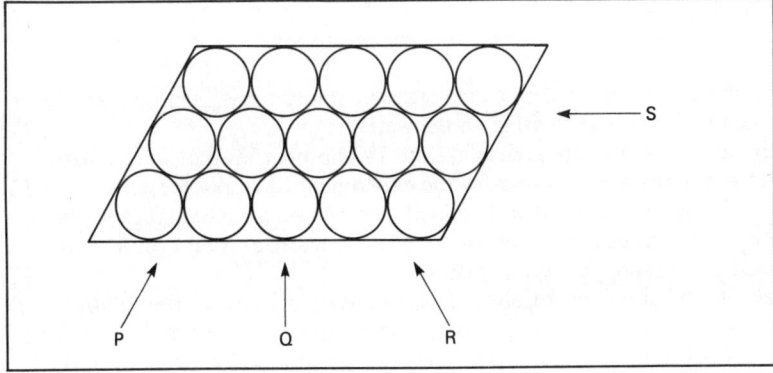

In which direction (s), P, Q, R or S, would you expect to be able to cleave this crystal? (There may be more than one direction possible.) Explain why you would expect the crystal to cleave in the way(s) you have suggested.   (2)

3. The Size of Olive Oil Molecules

A class of pupils was attempting to measure the size of a molecule of olive oil. In order to do this their teacher had prepared a solution of 0.1 cm³ of olive oil in 1000 cm³ of hexane. (Hexane is a liquid, boiling point 69°C, which resembles petrol in its properties). The solution was allowed to drip from a burette, and it was found that 1 cm³ of the solution gave (on average) 40 drops.

Each pupil then floated a loop of cotton on the surface of a bowl of water, and the hexane solution was dripped into it until the olive oil was just sufficient to fill the loop and keep it taut. The number of drops of solution needed just to fill the loop with olive oil was carefully counted. The area of the loop was also measured. The results of the class were plotted on the following graph.

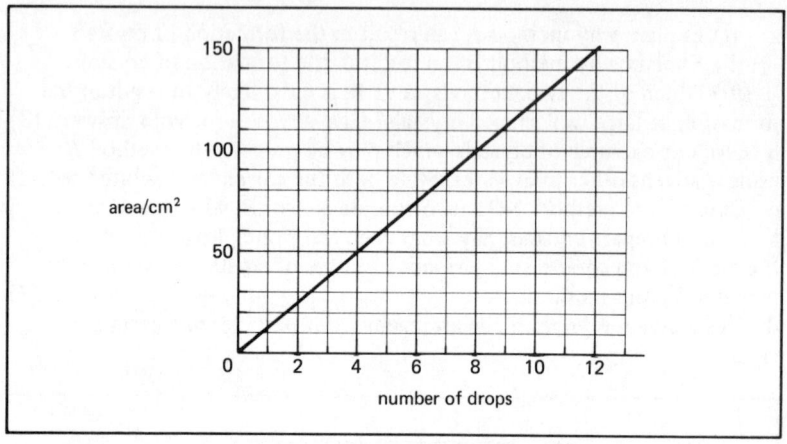

a What happens to the hexane once the solution has been dripped into the loop of cotton floating on the water?   (1)
b If no hexane were available, state TWO properties that an alternative solvent must have in order for the experiment to work successfully.   (2)
c Would the experiment described here be equally successful for finding the size of a sugar molecule of about the same size? State 'yes' or 'no', and give a reason for your answer.   (1)
d   (i) What volume of olive oil is contained in 1 cm³ of the solution?
    (ii) What volume of olive oil is contained in 10 drops of the solution?
    (iii) From the graph, determine the area covered by 10 drops of the solution.
    (iv) Use your answers to calculate the thickness of the layer of olive oil on the surface of the water, showing clearly your working.
    (v) State one assumption that must be made if this answer is to be taken as the size of an olive oil molecule.   (6)

4. Diffusion of Gases

Gases do not all diffuse at the same speed. The rate at which they diffuse depends upon their density, which, in turn, depends upon their relative molecular mass. This may be investigated by the following method:
    A glass syringe is clamped vertically and exactly 100 cm³ of the gas under investigation is drawn into it. A tube packed with glass wool is attached to the end of the syringe. The syringe plunger is allowed to fall

under its own weight, driving the gas out of the syringe and through the tube containing the glass wool.

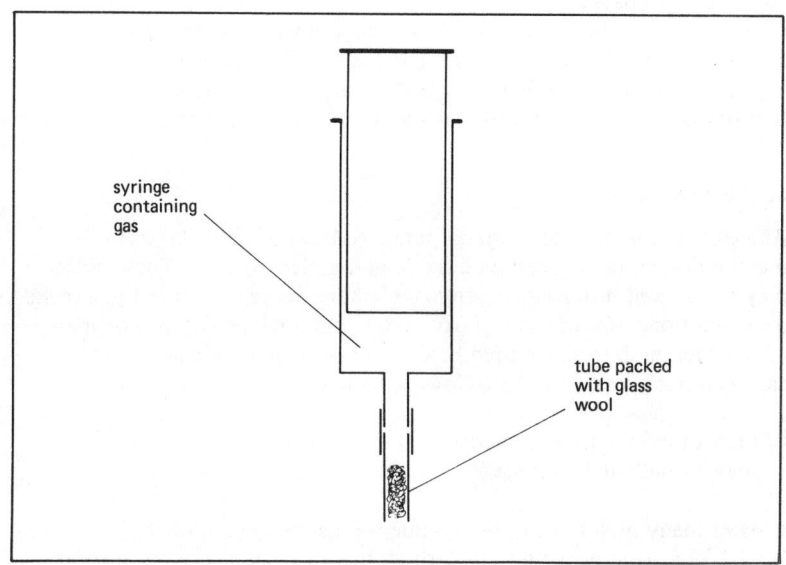

syringe containing gas

tube packed with glass wool

The time taken for 100 cm³ of the gas to escape in this way is noted down and compared for different gases. Typical results are shown in the following table:

Formula of gas	H_2	NH_3	O_2	HCl	CO_2	Cl_2
Mass of 1 mole/g	2	17	32	36·5	44	71
Time for 100 cm³ of gas to escape/s	5·5	16·0	21·9	23·3	25·7	32·6

Use the above information in answering the questions which follow.

a Explain what you understand by the term 'diffusion'. (1)
b At room temperature hydrogen molecules have a speed of about 1340 ms⁻¹ (3000 mph). Why, then, does it take 5·5 seconds for the hydrogen to diffuse out of the syringe? (1)
c Using the above data, plot a graph showing time for the gas to escape (on the horizontal axis) against the mass of 1 mole of the gas (on the vertical axis). (3)
d (i) North sea gas is methane, CH_4. Use your graph to estimate how long 100 cm³ of this gas would take to diffuse from the syringe.
(ii) 100 cm³ of air diffuse from the syringe in 20·6 s. What figure does this give on the graph for the mass of 1 mole of air (i.e. for the mass of air which occupies a volume of 24 000 cm³ at room temperature and atmospheric pressure). ? Explain how this value arises. (3)

e A child's balloon which is filled with hydrogen gets smaller more rapidly than one filled with air at the same temperature and pressure. Explain why this is so. (1)

f Suppose that the syringe used in the experiment described above were surrounded by a heating jacket kept at a steady 120°C. How would you expect the graph of the time against the mass of one mole to differ from that drawn in part (c)? Explain how you arrive at your answer. (1)

5. The Gas Laws

A device for making soda water at home consists of small disposable metal bulbs containing carbon dioxide under high pressure. These bulbs may be screwed into a larger container, where the gas is allowed to expand into about one litre of water. In an investigation of this device, the mass of one such bulb was measured before and again after discharging the carbon dioxide gas, with the following results:

mass of bulb before discharge = 26·98 g
mass of bulb after discharge   = 21·48 g

a How many moles of carbon dioxide were contained in the bulb? (2)

b (i) What volume would this carbon dioxide occupy as a gas at room temperature and atmospheric pressure.

(ii) Assuming that the volume of the bulb is 10 cm^3, what would be the pressure of the carbon dioxide if it were present in the bulb as a gas under these conditions?

(iii) Again assuming that the carbon dioxide is present as a gas, calculate to what figure the pressure would increase if the bulb were dropped in boiling water. (In this calculation, take room temperature to be 27°C.) (3)

c In actual fact, most of the carbon dioxide in the bulb is present as a liquid and not as a gas.

(i) Suggest a reason why the carbon dioxide is liquid, even at room temperature.

(ii) When the bulb is opened to allow the carbon dioxide to escape into the water, the bulb feels very cold. Why does the temperature drop, and what name is given to this change? (3)

d (i) Assuming that all the carbon dioxide dissolves in the water to produce one litre of aqueous solution, calculate the molarity of the solution formed.

(ii) A book of data states that at room temperature and atmospheric pressure, a saturated solution of carbon dioxide has a molarity of about 0·03 M. Suggest a reason why the solution prepared in this experiment has a molarity greater than 0·03 M. (2)

Section 2
Atomic Structure

6. Atomic Structure

Element gallium (Ga) exists naturally as a mixture of two isotopes, which may be represented by the symbols $_{31}^{69}$ Ga and $_{31}^{71}$ Ga. The lower figure is the atomic number and the upper one the relative atomic mass. 60% of the atoms are of the lighter isotope.

a (i) Explain what you understand by the term 'isotope'.
   (ii) Give the name of a gaseous element which exists as a mixture of two isotopes in the ratio of 3:1.   (2)
b Copy the following table and fill in the blanks.

	Number of protons per atom	Number of neutrons per atom	Number of electrons per atom
$_{31}^{69}$ Ga			
$_{31}^{71}$ Ga			

                                                                          (3)

c Calculate the relative atomic mass of naturally occurring gallium to one place of decimals, showing how you arrive at the answer.   (2)
d Gallium is similar to aluminium which forms an oxide, Al_2O_3.
   (i) Write the symbol for the gallium ion.
   (ii) Write a formula for the nitrate and the sulphate of gallium respectively.   (3)

7. Electrons and Structure

One atom of hydrogen will combine with one atom of oxygen to form a group which may be represented as shown below, where the outer or valency electrons are shown by dots.

$$\ddot{O}\!:\!H$$

a What is the name given to this group?   (1)

b The electron configuration of sodium is 2,8,1.

(i) Show, by a diagram, how the OH group would combine with one atom of sodium.

(ii) Give ONE chemical property of the compound formed in (i). (3)

c The electron configuration of hydrogen is 1 and of carbon is 2,4.

(i) Show, by a diagram, how the OH group would combine with a methyl group, CH_3.

(ii) Give the name of the compound formed. (3)

d The OH group is also present in the compound which is represented by the formula CH_3-C-OH (or CH_3COOH).

$$\overset{\|}{O}$$

(i) Give the name of this compound.

(ii) Which substance is formed, other than water, when this compound reacts with the one that you have described in (b)?

(iii) Which substance is formed, other than water, when this compound reacts with the one that you have described in (c)? (3)

8. A Metal like Calcium

The element strontium (Sr) is placed below calcium in group II of the periodic table. The atomic number of strontium is 38 and its relative atomic mass is 88.

a How many electrons are present in the outer shell of a strontium atom? (1)

b If the electron configuration of the element of atomic number 35 is 2,8,18,7, what would you expect that of an atom of strontium to be? (1)

c What is the composition of the nucleus of an atom of strontium? (2)

d Would you expect the combination of strontium with chlorine to result from

(i) sharing of electrons between atoms of strontium and chlorine,

(ii) transfer of electrons from atoms of strontium to atoms of chlorine,

(iii) transfer of electrons from atoms of chlorine to atoms of strontium? (1)

e How many moles of electrons are involved in the combination of 88 g of strontium with chlorine? (1)

f How many faradays of electricity are needed to discharge 1 mole of strontium ions during electrolysis? (1)

g Strontium reacts with water.

(i) Write down the names of the products that you would expect to be formed in this reaction.

(ii) Would you expect the rate of this reaction to be slower than, faster than, or about the same as that of calcium with water under the same conditions? Give one reason for your answer. (3)

9. Electronic Structures

This question is concerned with fluorine, neon and sodium, the atomic numbers of which are 9, 10 and 11, respectively.

a In each case state the number of protons present in the nuclei of fluorine, neon and sodium atoms.　　　　　　　　　　　　　(1)
b (i) In each case state how the electrons are arranged in atoms of fluorine, neon and sodium?
　(ii) State how the electrons are arranged in F^- and Na^+ ions. What do you notice about the arrangement?　　　　　　　　　(4)
c By means of suitable diagrams show the electrons are arranged in
　(i) an F_2 molecule
　(ii) sodium fluoride, $Na^+ F^-$.　　　　　　　　　　　　　(2)
d State one way in which you would expect the properties of the fluorine molecule to differ from those of sodium fluoride.　　　　　(1)
e Sodium atoms are very different from sodium ions in their behaviour.
　(i) Describe one example of differing behaviour which supports this statement.
　(ii) Suggest the cause of this different behaviour.　　　　　(2)

Section 3

The Mole Concept

---

10. Analysis of a Hydrocarbon

2.90 g of a dry gaseous hydrocarbon **Z** are completely burnt in excess of dry oxygen. The products are passed through a suitable drying apparatus containing anhydrous calcium chloride, and it is found that 4·50 g of water have been formed.

a (i) Calculate the mass of hydrogen in 4·50 g of water.
　(ii) Calculate the mass of carbon in 2·90 g of **Z**.
　(iii) Find the simplest possible formula for **Z**.　　　　　(4)
b (i) 100 cm^3 of Z were found to have a mass of 0·24 g at room temperature and atmospheric pressure. What is the relative molecular mass of **Z**?
　(ii) Find the molecular formula of **Z**.　　　　　　　　　(4)
c Another hydrocarbon **Y** has a molecular formula $C_4 H_8$. Write structural formulae for two possible isomers of **Y**.　　　　　(2)

11. Formula Determination

A colourless liquid **Q** boils at 46°C and burns readily in air. It contains only the elements carbon and sulphur, its mass being in the proportions of 15·8% carbon and 84·2% sulphur.

a What is the ratio of moles of carbon atoms to moles of sulphur atoms in **Q**? (3)

b If 1 mole of **Q** has a mass of 76 g, what is the molecular formula of **Q**? (2)

c Give the name of the TWO products that you would expect to be formed if some **Q** was burned in a plentiful supply of air. (2)

d Give the name of ONE product, which is different from those in part (c), that you would expect to be formed if the supply of air was limited. (1)

e When 1 mole of **Q** is formed from its elements, ΔH = +80 kJ per mole. Is this reaction exothermic or endothermic? (1)

f Suggest ONE precaution to take when storing this liquid in the laboratory. (1)

12. A Gas and its Formula

a (i) Explain why a gas exerts pressure on the walls of a closed vessel containing it.

(ii) How and why is this pressure affected by increasing the temperature of the gas? (4)

b 30 cm^3 of a gaseous oxide of carbon, measured at room temperature and atmospheric pressure, have a mass of 0·035 g.

(i) What is the mass of 1 mole of this oxide?

(ii) Use this information to suggest the most probable formula of the oxide. (4)

c At room temperature and atmospheric pressure, 1 g of hydrogen occupies 12 000 cm^3. Use this information to determine the atomicity of hydrogen. (2)

13. Avogadro's Law and Molar Volume

One mole of molecules of any gas always contains the same constant number of molecules. If the vapour density of a gas relative to hydrogen is known, the relative molecular mass can be found by doubling this value. From a knowledge of Avogadro's hypothesis we know that if we take equal numbers of molecules of, for example, oxygen, carbon dioxide and sulphur dioxide, then these gases will all occupy the same volume under the same conditions. The volume occupied by one mole of molecules of a gas is called the molar volume. In answering the following questions make use of the table of relative atomic masses given at the beginning of the book: the mass of one dm^3 of hydrogen at room temperature and atmospheric pressure is 0.0833 g.)

a (i) What name is used to describe the number of molecules in one mole of a gas?

(ii) How is this number affected by changes in physical conditions, e.g. in temperature and pressure? (2)

b (i) Why is the relative molecular mass of a gas twice its relative vapour density?

(ii) What is the relative vapour density of oxygen? (2)

c Air has a relative vapour density of 14·4: name ALL the gases from the following list which are LESS dense than air.

Methane   carbon dioxide   oxygen   ammonia (1)

d (i) What does Avogadro's hypothesis actually state?

(ii) A gas is at 27°C when the pressure on it is doubled. To what higher temperature must it now be heated for the original volume to remain unchanged? (3)

e Work out, from the figures given, the molar volume of hydrogen at room temperature and atmospheric pressure. (2)

14. The Formula of an Oxide of Copper

The diagram shows an apparatus which may be used to find the formula of an oxide of copper.

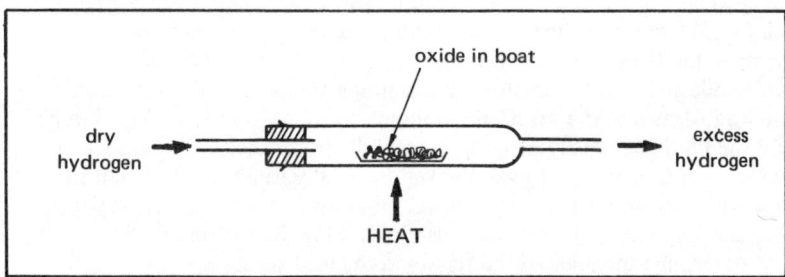

The mass of the boat was determined, first when it was empty, and then again with the oxide in it. Dry hydrogen gas was passed over the oxide which was heated. As it was being heated, and while the hydrogen was passing over it, a red glow was seen to spread through the oxide. When no further change seemed to be taking place, the heat was removed, but hydrogen was allowed to continue passing through the tube until it was cool. Finally, the boat – now containing copper – again had its mass determined.

mass of empty boat   = 11·2 g
mass of boat + oxide   = 14·8 g
mass of boat + copper   = 14·4 g

a What does the spread of the red glow indicate? (1)

b (i) What product is formed in this reaction other than the copper?

(ii) Why does the other product not interfere in the determination of the final mass? (2)

c How could the excess hydrogen be disposed of *safely*? (1)

d Why was it necessary to continue passing hydrogen over the copper until it was cool? (1)

e Give the name of one other oxide whose formula could be determined in an identical manner, and say why the method would be equally suitable. (1)

f (i) Calculate the mass of copper formed, and the mass of oxygen with which it was originally combined.

(ii) Calculate the number of moles of copper atoms formed, and the number of moles of oxygen atoms that were originally combined with them.

(iii) State the formula of the oxide.

(iv) Write a balanced equation for the reaction between this oxide and hydrogen. (4)

15. The Formula of Lead Bromide

The following experimental details were issued to a class which was attempting to determine the formula of lead bromide:

'Determine accurately the mass of a clean dry conical flask. Put into it about 1 g of lead bromide and again determine accurately its mass. Add about 50 cm^3 of pure water and heat to boiling. Remove the flask from the heat; add about 0·1 g of aluminium powder and boil gently for ten minutes. During this time the aluminium reacts with the lead bromide. Remove the flask from the heat and add about 20 cm^3 of dilute sodium hydroxide solution to dissolve the remaining aluminium. By this stage, the lead formed in the reaction has joined together to form a single lump and the liquid can easily be poured away. Wash the lead with pure water twice, each time pouring away the water. Finally, rinse the flask and the lead with propanone (acetone). Again, pour away the liquid before drying the flask by heating it very cautiously over a low Bunsen flame. When dry, determine the mass of the flask and the lead accurately.'

a (i) What are the products of the reaction between lead bromide and aluminium?

(ii) What does the experiment tell you about the relative positions of aluminium and lead in the reactivity (electrochemical) series? (2)

b Most of the masses had to be determined accurately: why was it satisfactory only to use *about* 0·1 g of aluminium powder? (1)

c (i) What is the precise purpose of washing the lead with water at the end of the experiment?

(ii) What is the purpose of washing the lead with propanone (acetone) at the end of the experiment?

(iii) Suggest TWO reasons why the flask with the lead in it has to be heated cautiously at the end of the experiment. (4)

d The following results were obtained in the experiment:

mass of empty flask	=	25·22 g
mass of flask + lead bromide	=	26·32 g
mass of flask + lead	=	25·84 g

(i) What mass of lead was produced in the reaction, and how many moles of lead atoms is this?

(ii) What mass of bromine was combined with this mass of lead, and how many moles of bromine atoms is this?

(iii) What is the formula for lead bromide? (3)

16. An Investigation of Iodine Trichloride

The following extract was taken from a pupil's laboratory notebook:
'In order to prepare some iodine trichloride I needed to generate a steady stream of chlorine gas. I did this by adding concentrated hydrochloric acid, a little at a time, to crystals of potassium permanganate. Chlorine was produced without needing to apply heat, and it was then passed through a U-tube containing iodine crystals. The crystals at first turned into a brown liquid, but later a yellow solid was formed all over the inside of the U-tube. I assumed that the yellow solid was iodine trichloride. There was a strong smell of chlorine. When nothing further appeared to be happening in the U-tube I disconnected the chlorine supply, weighed the U-tube and its contents, and then left it, uncorked, in a fume cupboard. When I came back several hours later the yellow solid had disappeared and had been completely replaced by the brown liquid. The smell of chlorine had gone completely from the U-tube, which now weighed less than it did before. I assume that the yellow solid had decomposed to give the brown liquid and chlorine, and thus we can write:

brown liquid + chlorine ⇌ yellow solid'

a Draw a labelled diagram of the apparatus you would use for this experiment (including the chlorine generator). (1)
b Explain the meaning of the sign ⇌ which appears in the equation at the end of the student's account. (1)
c Suggest and explain an experiment which the student could carry out in order to test whether the concluding equation is, in fact, correct. (2)
d The student claimed that chlorine could be smelt, in the U-tube, in the early part of the experiment, when the yellow solid was there, but not later on. By what process would the chlorine have been removed? (1)
e If the student's ideas are correct, how should she proceed in order to be able to keep a sample of the yellow solid without its turning into the brown liquid? Give a reason for your answer. (2)
f In a repeat of the experiment the following results were obtained:

mass of empty U-tube	= 59·83 g
mass of U-tube + iodine crystals	= 62·37 g
mass of U-tube + brown liquid	= 63·08 g

(i) Assuming that the brown liquid is composed only of iodine and chlorine, use this information to work out its formula. Show your working.

(ii) Now write the equation for the formation of the brown liquid from its elements. (3)

17. The Formula for Ammonia

The formula for ammonia may be determined by decomposing it into its elements, nitrogen and hydrogen, and then removing the latter by passing the mixed gases over hot copper(II) oxide. A suitable apparatus is shown.

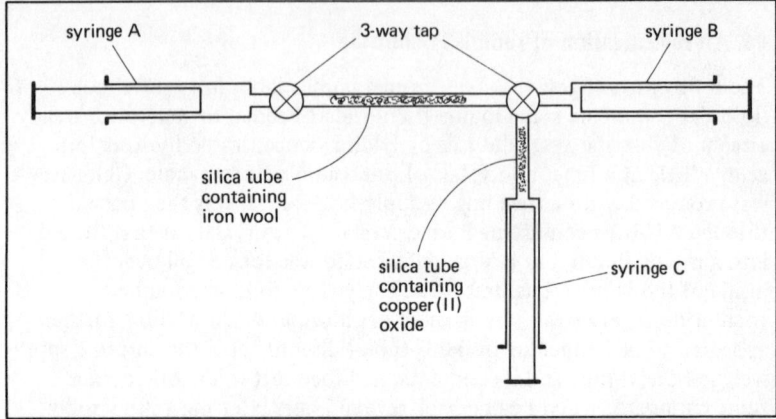

In an attempt to carry out the experiment at school, 40 cm³ of dry ammonia were passed into syringe B, syringes A and C being empty. The silica tube contained steel wool and was heated until red-hot, and the ammonia was then passed backwards and forwards several times between A and B. After some time, the heating was stopped, the apparatus allowed to cool to room temperature, and the remaining gas transferred back to syringe B. Its volume was now 60 cm³, and further passage through the hot steel wool did not alter this.

a   (i) What is the purpose of the steel wool?

    (ii) If the experiment has been successful so far, what should now be in syringe B?   (2)

The gas in syringe B was then passed several times between syringes B and C over hot copper(II) oxide, which changed colour from black to a pinkish brown. On cooling the apparatus to room temperature, a few drops of a clear liquid could be seen, and the volume of gas remaining was 20 cm³. Further passage of this gas over the hot copper (II) oxide produced no additional change.

b   (i) If the experiment has been successful, what gas remains now in the syringe?

    (ii) Explain why the copper(II) oxide changes colour and why a clear liquid is seen in the apparatus.

    (iii) What test would you carry out to confirm that you had correctly identified the liquid?   (3)

c Use the results of this experiment to calculate the volumes (at room temperature and atmospheric pressure) of the nitrogen and hydrogen produced by the decomposition of 40 cm³ of ammonia. Hence, calculate the *apparent* formula for ammonia. Show your working in full.   (3)

The formula for ammonia as determined in this experiment does not agree with that usually given. It was suggested that a possible reason for the error was that the steel wool was coated with a layer of iron oxide

d   (i) Why might a layer of iron oxide on the iron cause the results to be in error in this way?

(ii) Even if no other sample of steel wool were available for the experiment, suggest how you could modify the procedure so that this difficulty could be overcome.   (2)

18. A Reaction of Copper Ions

25 cm³ of a 0.1 M solution of copper(II) sulphate were taken and a dilute solution of ammonia was gradually run in. During the operation the titration vessel was continuously stirred to ensure a thorough mixing of the contents and to prevent the precipitate so formed from settling down. The amount of light transmitted by the mixture was measured on a suitable instrument and was related to the mass of precipitate present. The graph below shows the results.

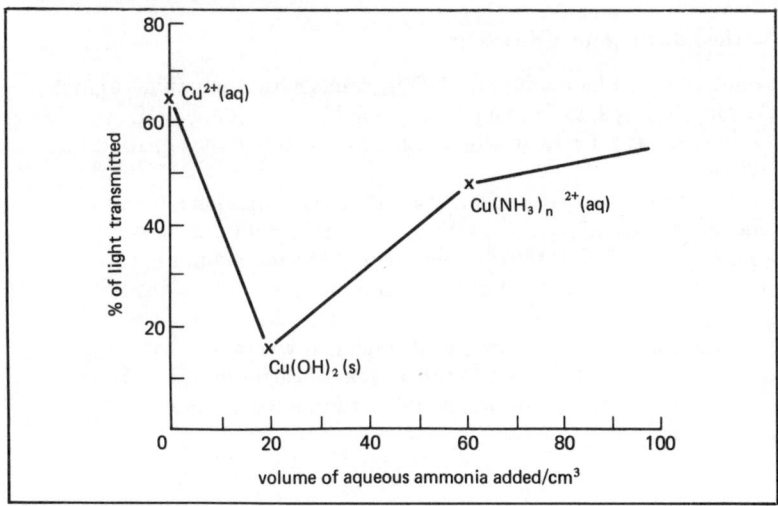

The percentage of light transmitted fell until 20 cm³ of ammonia solution had been added and then rose sharply until the added volume was 60 cm³ At this stage all the precipitate had redissolved and the further addition of the ammonia solution only caused a slight increase in the transmitted light. The two relevant equations are:

(1) $Cu^{2+} (aq)$ + $2OH^-(aq)$ $\longrightarrow$ $Cu(OH)_2 (s)$

(2) $Cu(OH)_2 (s)$ + $nNH_3 (aq)$ $\longrightarrow$ $Cu(NH_3)_n{}^{2+} (aq)$

15

a (i) How many moles of copper(II) sulphate are present in 25 cm³ of the 0·1 M solution?

(ii) Use equation (1) to determine how many moles of aqueous ammonia are needed to precipitate all this copper(II) sulphate as copper (II) hydroxide. (2)

b (i) What volume of aqueous ammonia contains this number of moles?

(ii) Calculate the number of moles of ammonia in 1 dm³ of the solution and hence determine the molarity. (2)

c (i) What additional volume of aqueous ammonia is needed to dissolve the precipitate completely?

(ii) How many moles of ammonia are contained in this volume? (2)

d (i) How many moles of copper(II) hydroxide are formed and subsequently redissolved?

(ii) Compare the number of moles of ammonia with the number of moles of coper(II) hydroxide dissolved by it and hence find the value of 'n'. (2)

e Re-write equation (2) with this value of 'n' and complete it. (1)

f Why does the percentage of light transmitted by the mixture still gradually increase after 60 cm³ of aqueous ammonia have been added? (1)

19. The Equation for a Reaction

Cadmium(II) sulphate solution, $CdSO_4$, reacts with ammonium sulphide solution, $(NH_4)_2 S$, to form a yellow precipitate of cadmium(II) suphaide. This question is concerned with an experiment to find the equation for this reaction.

0·5 mol cadmium(II) sulphate was dissolved in water and the solution made up to 500 cm³. 5 cm³ portions of the solution were measured into clean, dry test tubes. Different volumes of 2 M ammonium sulphide solution were added to the test tubes, and the precipitate formed at once. Each tube was shaken for 30 seconds and then placed in a centrifuge for 30 seconds. After centrifuging, it was found that the precipitates had settled down to the bottom of each tube. In each case the depth of the precipitate was noted, with the results shown in the following table.

Tube	Volume of $CdSO_4$/cm³	Volume of $(NH_4)_2 S$/cm³	Depth of precipitate/mm
A	5·0	1·0	7·0
B	5·0	2·0	14·5
C	5·0	3·0	18·0
D	5·0	4·0	17·0
E	5·0	5·0	19·0

a (i) What piece of equipment would you use to measure the volumes of ammonium sulphide solution?

(ii) Why was the same volume of cadmium(II) sulphate solution used in each experiment?

(iii) Why was it necessary to shake and centrifuge each mixture for the same length of time?

(iv) Suppose the centrifuge had been broken. Would it still be possible to carry out this experiment? Explain how you arrive at your answer and mention any differences you might expect if a centrifuge were not used. (4)

b Plot a graph of the results of this experiment, putting the height of the precipitate on the vertical axis and the volume of ammonium sulphide solution on the horizontal axis. (2)

c (i) What volume of ammonium sulphide solution is sufficient just to react with 5 cm^3 of the cadmium(II) sulphate solution?

(ii) How many moles of ammonium sulphide are present in this volume of solution?

(iii) How many moles of cadmium(II) sulphate are present in 5 cm^3 of the solution?

(iv) Write an equation for the reaction which takes place. (4)

Section 4
Structure and Bonding

20. Structures and Particles

This diagram summarises the types of particles found in chemical substances, and the structures which they form.

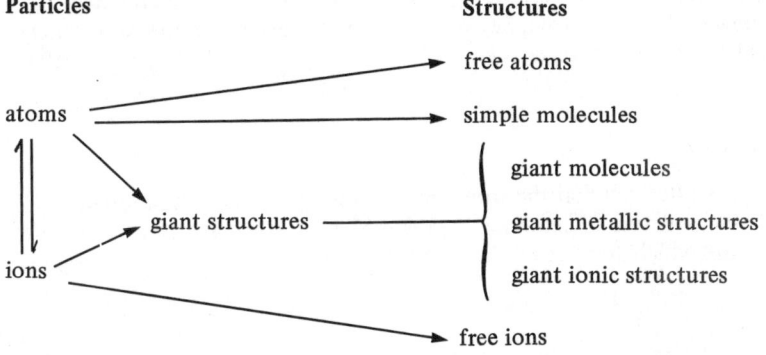

Use the summary in answering the questions which follow. You may need the information in the following table.

Substance	Formula	Melting pt./°C	Boiling pt./°C	Conducts electricity solid	molten
beryllium fluoride	BeF_2	800	1330	no	yes
boron trifluoride	BF_3	−129	−99	no	no
lanthanum	La	920	3367	yes	yes
magnesium oxide	MgO	2900	3600	no	yes
sodium chloride	NaCl	808	1465	no	yes

a (i) The table gives some information about melting and boiling points. What may be deduced about the likely structure of a substance having low melting and boiling points?

(ii) The table also gives some information about the electrical conductivity of the solid substances. Why is this useful in deciding about structures?

(iii) What extra information can be derived from the electrical conductivity when molten? (3)

b Using the information in the table, state what the likely structure will be for each of the following:

(i) lanthanum (ii) beryllium fluoride (iii) boron trifluoride (3)

c A schoolboy, in answering a question which was set for homework, copied the following statement from an old chemistry book: 'Sodium present in the sodium chloride molecules is replaced by hydrogen from the sulphuric acid.' Explain why his teacher marked it wrong. (1)

d State one way by which a giant structure of ions can be converted into a system of free ions. (1)

e (i) From your own knowledge, (i.e. not necessarily using substances referred to in the table) give an example of how an atom may be converted into an ion. Say what atom you are using and how you would carry out the conversion.

(ii) From your own knowledge, give an example of how an ion may be converted into an atom. Say what ion you are using and how you would carry out the conversion. (2)

21. Carbon

Chemists often say that the structure of carbon in graphite is different from the structure of carbon in diamond both of which are shown in the diagrams, which are drawn to scale.

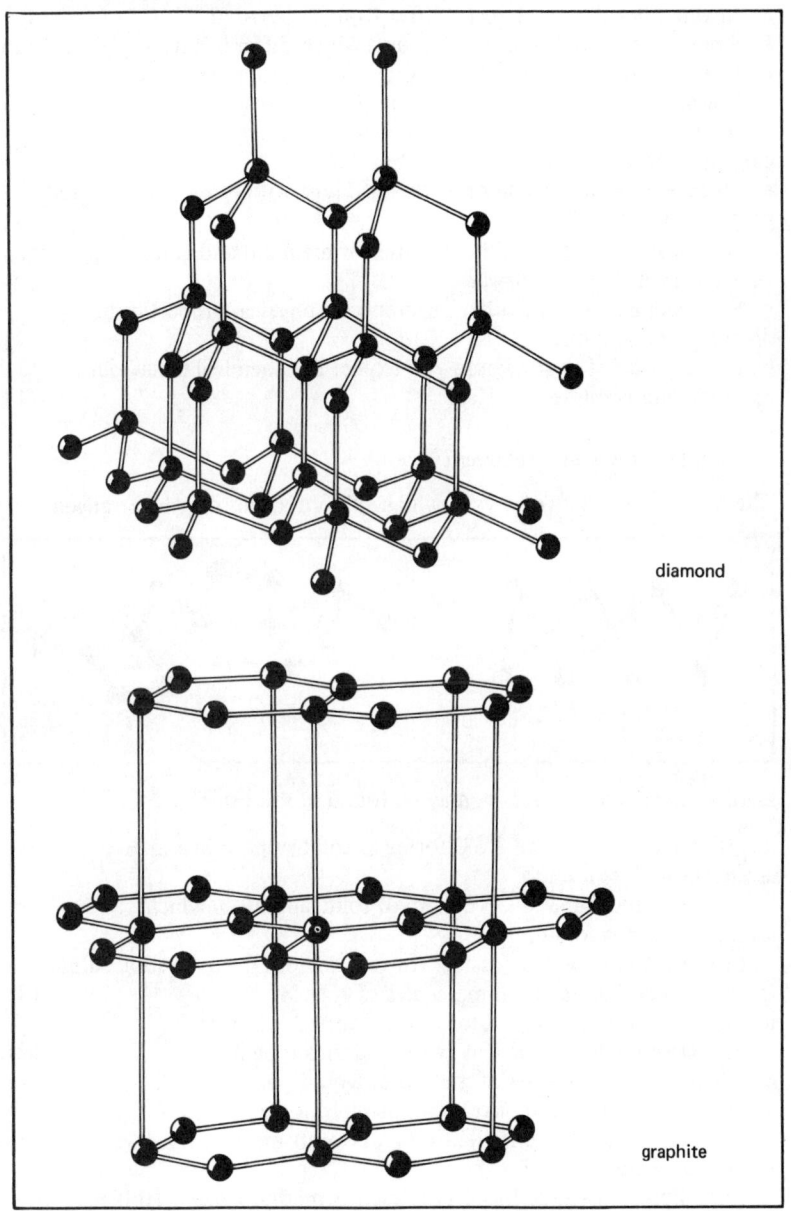

diamond

graphite

Equally, they may say that carbon-12($^{12}_6$C) differs from carbon-13($^{13}_6$C).

a In what way does carbon-12 differ from carbon-13? (1)
b What word is used to describe atoms which differ in this way? (1)
c How would you expect the densities of solid samples of carbon-12 and carbon-13 to differ? Give a reason for your answer. (2)
d In what way, if at all, would you expect the chemical behaviour of carbon-12 to differ from that of carbon-13? (1)
e What word is used to describe the different structures, diamond and graphite, of carbon? (1)
f How would you expect the densities of diamond and graphite to differ? Give a reason for your answer. (2)
g State and explain one other difference in physical properties between diamond and graphite. (1)
h In what way, if at all, would you expect the chemical behaviour of diamond and graphite to differ? (1)

22. Sulphur, the yellow element

The diagrams show two ways in which sulphur atoms may be arranged.

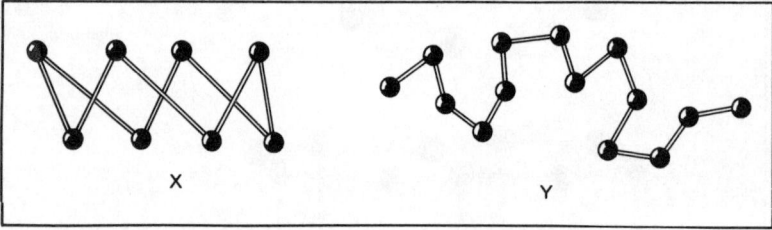

Both arrangements, X and Y, may be found in solid sulphur.

a (i) Give the names of TWO forms of solid sulphur in which arrangement X is found.

(ii) Give the name of ONE form of solid sulphur in which arrangement Y is found.

(iii) Explain how it is possible for two distinct forms of solid sulphur to exist which *both* contain molecules of type X. (3)

b (i) Describe how type X may be converted into type Y.

(ii) Describe how type Y is converted into type X. (2)

c Which of the two types of molecules would you expect in
(i) sulphur heated to *just* above its melting point
(ii) sulphur heated sufficiently far above its melting point for the colour to be almost black.
(iii) sulphur dissolved in solvents such as methylbenzene (toluene) or dimethylbenzene (xylene)? (3)

d (i) How would you expect the density of type Y (in the solid state) to differ from that of type X (also in the solid state)? Explain how you arrive at your answer.

(ii) State one other difference in the physical properties of solid X and Y molecules and explain how this difference arises. (2)

23. The Shape of Hydrides and their Formulae

The table below shows the formulae of some of the compounds with hydrogen (hydrides) formed by elements in the two periods of the periodic table which come after hydrogen and helium. In addition, four elements are denoted in the table by the symbols **W, X, Y** and **Z**. You are NOT expected to identify these elements.

Group	1	2	3	4	5	6	7	0
Element	Li	**W**	**X**	C	N	O	F	Ne
Hydride	LiH			CH_4	NH_3	OH_2	FH	–
Element	Na			**Y**	**Z**	S	Cl	Ar
Hydride	NaH					SH_2	ClH	–

a Write down the formulae for the hydrides of **W, X, Y** and **Z**, using these letters to represent the elements. (2)

b Lithium hydride is an ionic compound, $Li^+ H^-$. If molten lithium hydride were electrolysed with inert electrodes, state what you would expect to be liberated

(i) at the positive electrode (anode),

(ii) at the negative electrode (cathode). (2)

c From three different groups choose THREE hydrides which exist as simple molecules and draw diagrams to represent the shape of each molecule that you choose. (3)

d Why does neon (Ne) not form a hydride? (1)

e Give the formula of ONE product that you would expect to be formed if the hydride of **Y** was burned in a plentiful supply of air. (1)

f How would you expect the hydride of **Z** to affect moist pH or litmus paper? (1)

24. The Structure of some Chlorides

This table gives some information about the first ten elements in the periodic table and the compounds which they form with chlorine.

Symbol	H	He	Li	Be	B	C	N	O	F	Ne
Atomic number	1	2	3	4	5	6	7	8	9	10
Formula of chloride	HCl	–	LiCl	$BeCl_2$	BCl_3	CCl_4	NCl_3	OCl_2	FCl	–
Melting point of chloride/°C	-114	–	614	410	-107	-23	-37	-20	-154	–
Boiling point of chloride/°C	-85	–	1382	547	13	77	71	2	-101	–

atomic number of chlorine = 17

a Why are no data given for the chlorides of helium and neon? (1)

b Which of the chlorides listed would be liquid at room temperature? (1)

c (i) What evidence is given in the table to suggest that oxygen chloride probably consists of molecules?

(ii) Draw a simple 'dot-and-cross' diagram to show how the outer electrons are likely to be arranged in a molecule of oxygen chloride. (3)

d (i) From the chlorides listed in the table, give the formula of ONE which is likely to be composed of ions, stating what evidence there is to support your choice.

(ii) By means of suitable 'dot-and-cross' diagrams show how your chosen ionic chloride is derived from its constituent atoms. (3)

e Boron trichloride, BCl_3, can be made by heating solid boron in a stream of dry chlorine gas, although the product is rapidly decomposed by water, even in the cold. Assuming that you had a supply of chlorine available, draw a labelled diagram of the apparatus you would use to prepare and collect a reasonably pure sample of boron trichloride. (2)

25. The Different Structures of Substances

The diagrams at the right represent a number of different structures which substances may assume.

a (i) What type of structure is represented by figure I?

(ii) Give the name of one substance which you would expect to have the structure shown in figure I.

(iii) How could the structure shown in figure I be converted into that shown in figure II?

(iv) How would you expect the density of the structure represented by figure I to differ from that of figure II? Explain how you arrive at your answer. (4)

b (i) What type of structure is represented by figure III?

(ii) Give an example of a substance which would have this structure *at room temperature.*

(iii) State TWO ways in which the properties of a substance having the structure shown in figure III would differ from those of a substance with the structure shown in figure I.

(iv) By what method could the structure shown in figure III be converted into that shown in figure IV? (5)

c The structure shown in figure V could represent that of all of the members of one 'family' of chemicals at room temperature. What 'family' is this? (1)

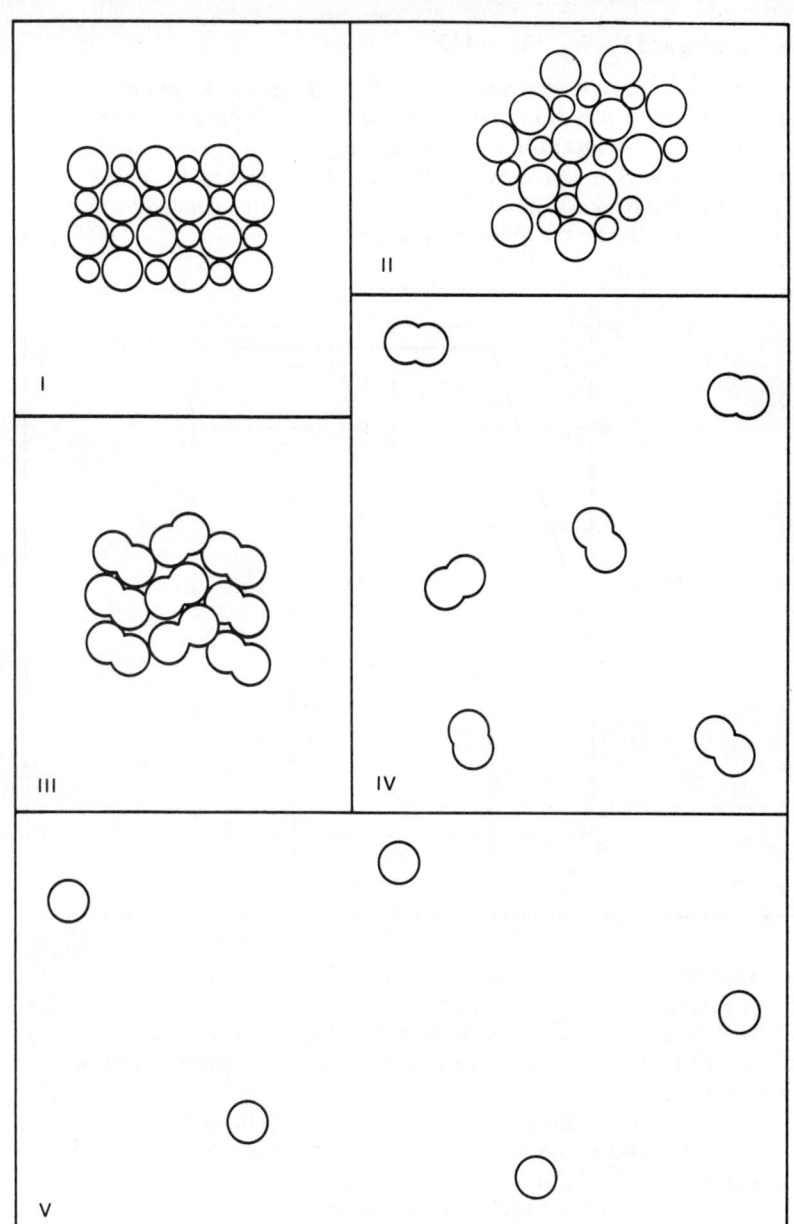

26. Heating and Boiling 'Freon-113'

A coil of wire was connected to a d.c. supply of electricity, and to
suitable meters which showed that the coil was supplying a constant
amount of heat at 1200 J min^{-1}. The heating coil was dipped into 93.75 g
of a pure liquid chemical known as 'freon-113', which has the formula
$C_2F_3Cl_3$. The temperature of the freon-113 was noted regularly and
the results plotted on the following graph.

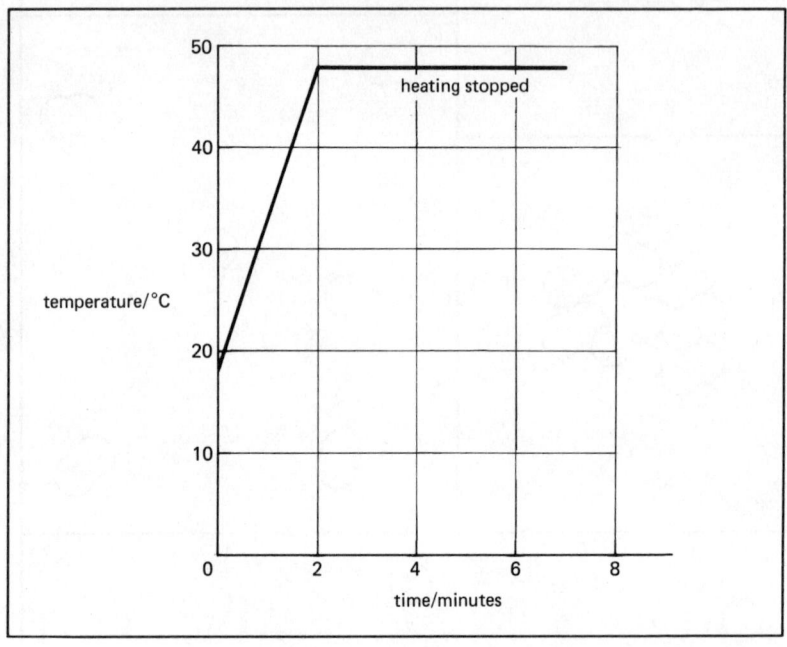

a What was room temperature on the day of the experiment? (1)
b What is the boiling point of freon-113? (1)
c What type of structure would you expect freon-113 to have? (1)
d How many moles of freon-113 were used in the experiment? (Show
your working.) (1)
e (i) How long did the freon-113 take to reach its boiling point?
(ii) How much energy is required to raise the temperature of 1 mole
of freon-113 by 1°C? (Show your working.) (3)
f (i) For how long was the freon-113 boiling?
(ii) In the time that the freon-113 was boiling, 46·88 g of it were
vaporised. How much energy is required to vaporise 1 mole of freon-113?
(3)

Section 5
Electrochemistry

27. Electrical Conduction

Questions (a) to (c) refer to the following list of substances:

diamond	phosphorus	potassium chloride
hydrogen chloride	sucrose	lead(II) bromide
sulphur	mercury	zinc

a (i) Which of the listed substances conduct electricity in the solid state?

(ii) Why are these substances, and only these, conductors when solid? (2)

b (i) Which of the listed substances dissolve in water to produce a conducting solution?

(ii) Explain why these solutions conduct electricity. (3)

c (i) Which of the listed substances conduct electricity when molten?

(ii) Substances which do conduct electricity when molten may be subdivided into two groups on the basis of their behaviour when conducting. In what way do these two groups behave differently? Illustrate your answer by explaining what happens with ONE example from EACH group. (5)

28. Products of Electrolysis

All the elements in the following list may be produced by the electrolysis of suitable molten substances or aqueous solutions.

bromine	oxygen	chlorine
silver	copper	sodium
hydrogen	zinc	lead

a (i) Explain what you understand by the term 'electrolysis'.

(ii) Why must substances be either molten or in aqueous solution in order that electrolysis may take place? (2)

b From the above list pick TWO elements whose ions would be discharged at the cathode, and for each one give the formula of the ion which would be discharged to produce that element. (2)

c From the above list pick TWO elements whose ions would be discharged at the anode, and for each one give the formula of the ion which would be discharged to produce that element. (2)

25

d One of the elements in the list can only be made by the electrolysis of a *molten* substance. State which element this is, and explain why an aqueous solution would not be suitable in this case. (1)

e The reactions which take place during the electrolysis of aqueous solutions can often vary according to the concentration of the solution or to the electrodes used. From your own knowledge (i.e. not necessarily) using elements from the above list) give one example where this is the case. You should give the name of the solution used, and state which two different reactions can take place under the two different conditions which you give. (3)

29. Electrolysis and Moles

A current of 2·00 amperes was passed for 1920 seconds through an aqueous solution of a salt of a metal **M**. The mass of **M** deposited on the negative electrode was 2·24 g.

a Would you expect the ions of the metal **M** to be positively charged or negatively charged? Give a reason for your answer. (1)

b How many coulombs were passed in the experiment described? (1)

c If the relative atomic mass of **M** is 112, how many moles of atoms of **M** are present in 2·24 g of **M**? (2)

d How many coulombs would be needed to deposit 1 mole of atoms of **M**? (2)

e 1 faraday (1 mole of electrons) is 96 000 coulombs. How many charges are carried by an ion of **M**? (2)

f Write an equation to show what happens to an ion of **M** when it is discharged at the negative electrode. (2)

30. Electrolysis of Copper Compounds

The diagram shows an electric circuit.

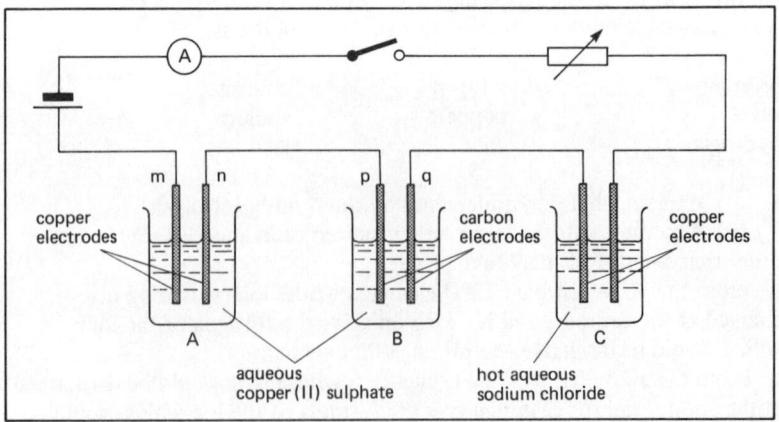

The current passes in turn through three cells, A, B, C, two of which contain aqueous copper(II) sulphate, and the third contains hot sodium chloride solution. Copper electrodes are used in A and C, and carbon ones in B. A constant current of 0.5 A was passed through the circuit for 10 minutes.

a What was the purpose of the variable resistor (rheostat)? (1)
b Copper is deposited on one electrode in each of the cells A and B.
  (i) On which electrodes, m, n, p, q, is this copper deposited?
  (ii) Without carrying out any calculation, what can you say about the mass of copper deposited on each of these two electrodes?
  (iii) In cell A the other copper electrode dissolves. How many moles of copper atoms would dissolve for the passage of 1 faraday of electricity (1 mole of electrons)? Explain how you arrive at your answer. (4)
c In cell C, the cathode did not change in mass during electrolysis, but the anode decreased in mass by 0·2 g. A reddish solid collected on the bottom of the beaker.
  (i) What quantity of electricity was used in the experiment?
  (ii) How many moles of copper atoms were lost by the anode?
  (iii) How many moles of copper atoms would have been lost from this electrode by the passage of one faraday of electricity (1 mole of electrons)?
  (iv) Assuming that all the copper which is lost from the anode ends up as copper ions in the red solid, what is the charge on those ions?
  (v) Write an equation for the process taking place at the anode. (5)

31. Electrolysis of a Melt

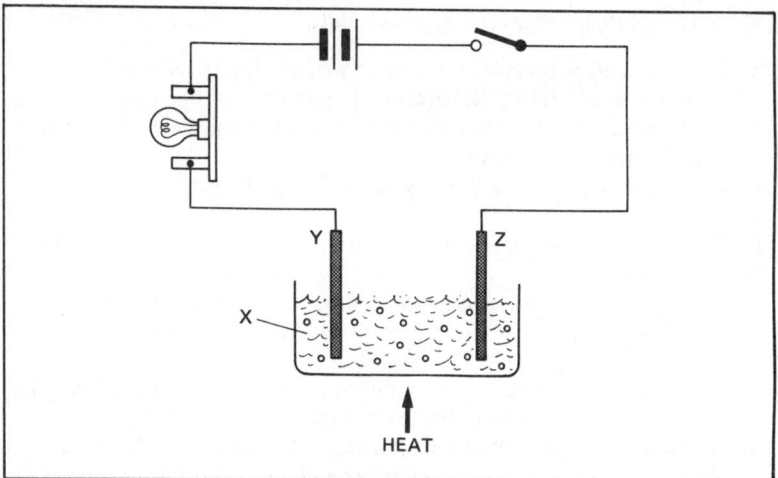

X represents a yellow compound into which two electrodes, Y and Z, are dipping. At the start of the experiment (at room temperature) the switch was closed, but the lamp did not light.

a   (i) Why did the bulb not light up when the switch was closed?

    (ii) What name is given to the electrode Y?

    (iii) What name is given to the electrode Z?

    (iv) Which is the positive electrode, Y or Z?

The basin was then heated until a temperature of 685°C was reached and the bulb lit up. Puffs of a violet vapour V came from around one electrode. (2)

b   (i) What element do you think V is likely to be?

    (ii) At which electrode, positive or negative, is V likely to appear? (2)

The electrode at which the violet vapour had NOT been given off was removed and examined and found to have a dull metallic-like coating, M When it was rubbed on a piece of paper it made a mark very much like that of a pencil.

c   (i) What element do you think M is and why?

    (ii) What kind of structure is X likely to have? Give a reason.

    (iii) What compound do you think X is?

    (iv) What is the formula of compound X if element M is divalent? (4)

The solubility of X is about 0.076 g in 1 dm^3 of water at room temperature, but about 10 times this amount at 100°C.

d Explain what you would notice if 0·10 g of X were shaken with 100 g of cold distilled water, boiled and then cooled under the tap. (2)

32. Electrolysis of Molten Metal Chlorides

Samples of sodium chloride and magnesium chloride were each placed in suitable containers and heated strongly until they melted. A pair of electrodes was placed in each and these were connected in series with an ammeter, a variable resistor (rheostat), and a source of d.c. electricity.

a If you wished to measure the quantity of electricity used in the experiment, what additional piece of equipment would be needed? (1)

b When choosing a suitable material to be used as the electrodes, state TWO properties it must have. (1)

c Why would it be unsafe to carry out this experiment in the open laboratory? (1)

d During the experiment a constant current of 5 A was allowed to flow for 576 seconds.

    (i) What quantity of electricity was used in the experiment?

    (ii) How many faraday (moles of electrons) were used in the experiment? (2)

e Socium and magnesium were produced at the cathodes. In each case the metal was carefully removed, cleaned free from excess chloride, and its mass determined. 0.69 g of sodium and 0.36 g of magnesium were obtained. (5)

    (i) Calculate the number of moles of sodium atoms formed during the experiment.

    (ii) Use the results of the experiment to calculate how many faradays are required to discharge 1 mole of sodium atoms. Show your working.

(iii) Calculate the number of moles of magnesium atoms formed during the experiment.

(iv) Use the results of the experiment to calculate how many faradays are required to discharge 1 mole of magnesium atoms. Show your working.

(v) Compare your answers to (e) (ii) and (iv), and explain why the answers obtained are different for the two metals.

33. Electrolysis helps Identification

A student was given three liquids, A, B and C, the first being coloured and the other two colourless. He was told to find out if they would conduct electricity, and then to carry out further tests to try to identify them as far as he could since complete identification might not be possible in all three cases.

He found that A and C, but not B, conducted electricity. The products of electrolysis are listed here:

	Cathode	Anode
Liquid A	red deposit	red-brown gas giving white fumes with ammonia
Liquid C	flammable gas	green gas which decolorised litmus solution

a What type of structure do you think is present in (i) liquid A, (ii) liquid B, (iii) liquid C? (1)

b Give the name and formula of the gas obtained from liquid C
(i) at the cathode   (ii) at the anode. (2)

Blue litmus paper dipped into B and C was unaffected by B, but turned red in C. Gentle evaporation of the three liquids was carried out with these results:

A coloured powder left
B no deposit; vapour caught fire
C no deposit; white fumes evolved

c State what class of compound (*not* type of structure) is possibly present in (i) liquid A, (ii) liquid C. (1)

d Give the name and formula of compound C. (1)

The powder obtained from liquid A by evaporation was melted between two carbon electrodes forming part of an electric circuit containing a bulb. Until the liquid had melted nothing happened, but then the bulb lit up. Puffs of a red-brown gas or vapour came off at the anode and the cathode again became covered with a reddish-brown deposit.

e (i) Give the names and formulae of TWO brown gases or vapours.

(ii) Suggest the name and formula of the anion present in liquid **A**. (2)

A. The reddish-brown deposit on the cathode was removed and placed in some nitric acid in a beaker in a fume cupboard. The liquid became green in colour and some dark brown fumes appeared above it; these fumes were NOT the same as those previously seen.

f (i) Identify the reddish-brown deposit.

(ii) Give the name and formula of compound **A**. (2)

Finally, **B** was mixed with some ethanoic (acetic) acid and a little concentrated sulphuric acid and warmed for a few minutes in a test tube. A sweet-smelling liquid was formed and the smell became more evident on pouring the warm mixture into cold water in a beaker.

g To what class of compounds does **B** belong? (1)

Section 6

Thermochemistry and Enthalpy

34. Hydrocarbon Fuels

The following table gives the name, formula and heat of combustion (ΔH_c) for a number of hydrocarbons. Use this information to answer questions (a) to (e).

Name	Formula	Heat of combustion /kJ mol^{-1}
methane	CH_4	-890
ethane	C_2H_6	-1560
propane	C_3H_8	-2220
butane	C_4H_{10}	-2877
pentane	C_5H_{12}	-3509
hexane	C_6H_{14}	-4195
heptane	?	?

a Predict the formula for heptane, the next member of this series. (1)

b What is the significance of the negative sign in the values of the heats of combustion? (1)

c (i) Draw a graph of ΔH_c (on the vertical axis) against the number of carbon atoms in each hydrocarbon (on the horizontal axis).

(ii) From this, predict ΔH_c for heptane. (3)

d Many of the hydrocarbons listed in the above table have important uses as fuels in everyday life; give TWO examples of such a use. (2)

e The excreta of pigs can be processed to produce approximately 275 dm³ of methane per pig per day, measured at a temperature of 37°C.

(i) How many moles of methane could be produced daily from one pig? (Assume that the molar volume of all gases at 1 atmosphere and 37°C is 25 dm³)

(ii) How much heat could be produced daily by one 'pig's-worth' of methane?

(iii) A typical home might require an average of 55 000 kJ per day for cooking and hot water. How many pigs would be needed to supply this demand? (3)

35. The Carbon Cycle

The following diagram, known as the carbon cycle, summarises the manner in which carbon compounds may be interconverted in nature, and with man's help.

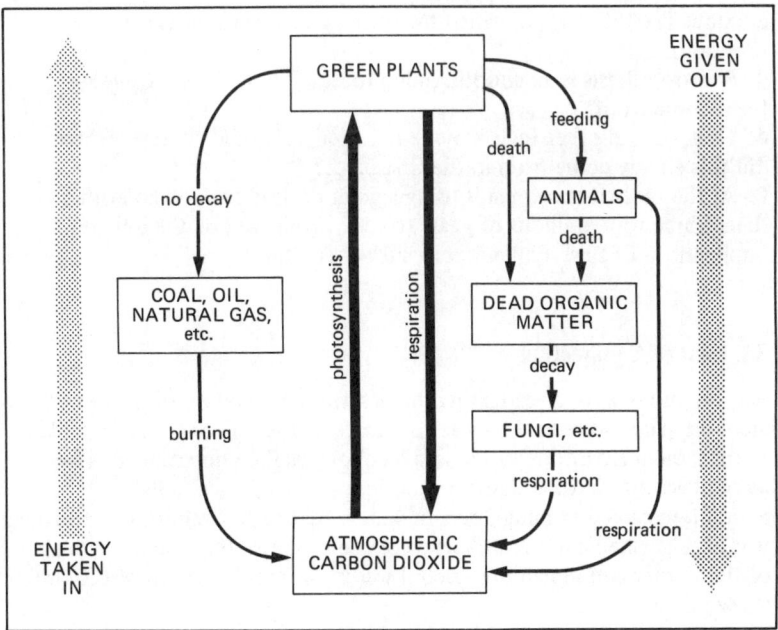

Photosynthesis is the process whereby large molecules (e.g. glucose, $C_6H_{12}O_6$) – are built up from carbon dioxide and water in the presence of sunlight. The equation for this may be written

$$6CO_2 + 6H_2O \longrightarrow C_6H_{12}O_6 + 6O_2$$

It is an endothermic process with ΔH = +1257 kJ/mol, and this may be represented on an energy level diagram such as the following:

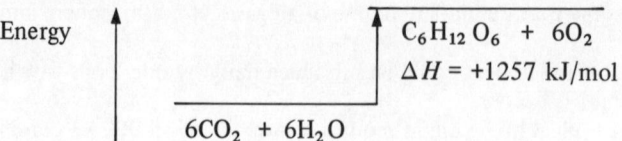

Energy

$$C_6H_{12}O_6 \ + \ 6O_2$$
$$\Delta H = +1257 \text{ kJ/mol}$$

$$6CO_2 \ + \ 6H_2O$$

Use this information in answering the questions which follow.

a (i) Respiration may be thought of as the reverse of photosynthesis. Write an equation to represent respiration.

(ii) Copy the energy level diagram shown above for photosynthesis, and draw the corresponding energy level diagram for respiration alongside it and to the same scale. (2)

b Natural gas is methane, CH_4. Write a balanced equation for its complete combustion in a plentiful supply of air. (2)

c State THREE ways in which the burning of fuels is similar to respiration. (3)

d Photosynthesis is an endothermic process. Where does the energy for it come from? (1)

e Coal, oil and other fuels may be regarded as 'stored energy'. Where did this energy come from in the first place? (1)

f At the present time man is using a great deal of coal, oil and other fuels which took millions of years to form. What will be the long-term implications of this? Explain your answer carefully. (1)

36. Heat of Combustion

Several pupils were investigating the heats of combustion of a series of alcohols. They were told to place a sample of their alcohol in a suitable burner; the mass of the burner and alcohol was then determined. The alcohol was lit, and the flame produced was used to heat a light aluminium can containing 200 g of water. After a few minutes, the flame was extinguished and the temperature rise of the water noted. The mass of the burner and remaining alcohol was then found. Typical results are shown here:

alcohol used: ethanol, C_2H_5OH

mass of burner and alcohol at the start	= 53·21 g
mass of burner and alcohol at the end	= 52·75 g
initial temperature of the water	= 19°C
final temperature of the water	= 27°C

Assume that 4 J raise the temperature of 1 g of water by 1°C.

a What quantity of heat did the alcohol supply to the water? Show your
working. (2)
b (i) How many moles of alcohol were burnt in the experiment?
(ii) Use this figure, and the result of (**a**) above, to obtain an estimate
of the heat produced by burning 1 mole of ethanol.
(iii) State what assumption you have to make in order to arrive at
this answer.
(iv) Would ΔH for the combustion of ethanol be positive or negative? (4)
c Another pupil in the class used propanol instead of ethanol. How
would you expect his value of ΔH to differ from that of the first pupil?
Explain how you arrive at your answer. (2)
d A further pupil in the class used butanol and observed a yellow flame,
which he attributed to the formation of carbon. How would this pupil's
value of ΔH differ from the value he would have obtained if no carbon
had been formed? Explain how you arrive at your answer. (2)

37. Energy changes on displacement

25·0 g of a powdered metal **X** are placed in 1 dm^3 (1 litre) of 1·0 **M** lead
nitrate solution and a dirty grey precipitate forms. The temperature rises
and 160 kJ of heat (ΔH) are liberated. 1.0 g of **X** remains unchanged.

a (i) Is the reaction exothermic or endothermic?
(ii) Is ΔH positive or negative? (2)
b (i) Is **X** above or below lead in the electrochemical series?
(ii) **X** has the same valency as lead has in its common compounds;
what is this valency? (2)
c From the figures given, calculate the relative atomic mass of **X**. (1)
d If 1 kJ of heat energy is sufficient to raise the temperature of 1 dm^3
of solution by 0·24°C, calculate, to the nearest degree, the theoretical
rise in temperature (i.e. ignore any heat losses). (1)
e Write the equation for the reaction between **Y** and the lead nitrate
solution. (1)
f (i) What mass of **X** would be used up if only 500 cm^3 of 1·0 **M**
lead nitrate solution is used?
(ii) How much heat would be evolved in this case?
(iii) What would be the temperature rise in this case (again, ignore
any heat losses). (3)

38. Calor Gas and The Gas Laws

Calor gas is mainly butane, C_4H_{10}, although there are some other hydrocarbons present in small amounts. For the purpose of this question we
shall consider it to be entirely butane.

a (i) Give the names of the products of combustion when Calor gas
burns in plenty of air.
(ii) Write the balanced equation for his reaction. (2)

b (i) What volume of oxygen (measured at the same temperature and pressure as the gas) would be needed for the complete combustion of 6 dm³ of Calor gas?

(ii) If the air contains 20% of oxygen by volume, what volume of air would be needed for this reaction?

(iii) What would be the mass of oxygen used in the reaction? (You may consider all volumes to be measured at room temperature and pressure, r.t.p.)   (3)

c (i) What mass of Calor gas would need to be burned in order to produce 13 500 kJ of heat? (Take the molar heat of combustion of butane as 3000 kJ.)

(ii) Would ΔH for this reaction be positive or negative?   (2)

d What would be the volume of 0·1 mole of butane at

(i) r.t.p. (20°C and 1 atmosphere pressure),

(ii) 20°C and two atmospheres pressure,

(iii) 313°C and two atmospheres pressure?   (3)

Section 7
Rates of Reaction

39. Rates of Reaction

The curves below show the results of two experiments which were carried out using magnesium ribbon and a dilute acid. In each case, the same mass of magnesium ribbon was taken and the same volume of acid of the same concentration was used. The acids were 1 M hydrochloric acid and 1 M ethanoic (acetic) acid.

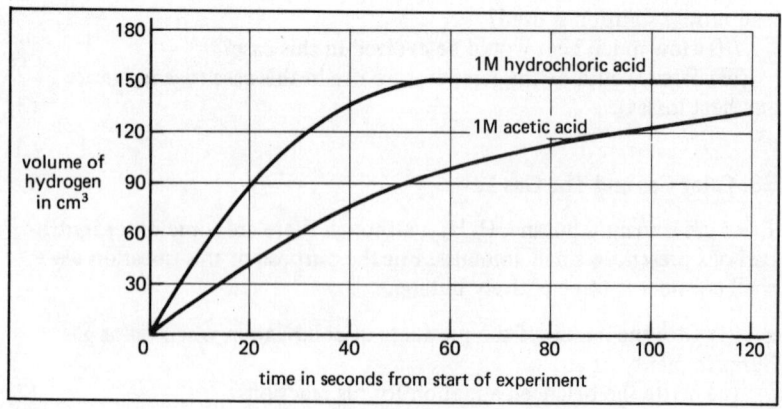

a An equation which represents *both* reactions is

$$Mg(s) + 2H^+ (aq) \longrightarrow Mg^{2+} (aq) + H_2(g)$$

(i) What do the curves tell you about the concentration of H^+ ions in each acid solution?

(ii) If left long enough, the volume of hydrogen collected at room temperature and pressure is the same in each case. Why is this so?　　(3)

b If the volume of hydrogen produced is 150 cm³ at room temperature and pressure, calculate the mass of magnesium used in each experiment.(2)

c How many seconds would elapse before half the magnesium was used up

(i) with hydrochloric acid,　　(ii) with ethanoic (acetic) acid?　　(2)

d Copy the graph and sketch on it the curve you would expect if the same mass of powdered magnesium were used with the same volume of 1 M hydrochloric acid under the same conditions.　　(1)

e In order to obtain a similar curve to that given by magnesium ribbon and 1 M hydrochloric acid, using the same mass of magnesium and the same volume of a fully ionised acid H_2A, what concentration of the acid in moles per dm³ would be used?　　(2)

40. Factors affecting the rate of reaction between Zinc and Hydrochloric Acid

When granulated zinc is added to excess dilute hydrochloric acid, hydrogen gas is produced. The speed of the reaction may be measured by collecting the hydrogen and measuring its volume at regular time intervals.

$$Zn + 2HCl \longrightarrow ZnCl_2 + H_2$$

The results obtained in one experiment are shown in curve A on the graph.

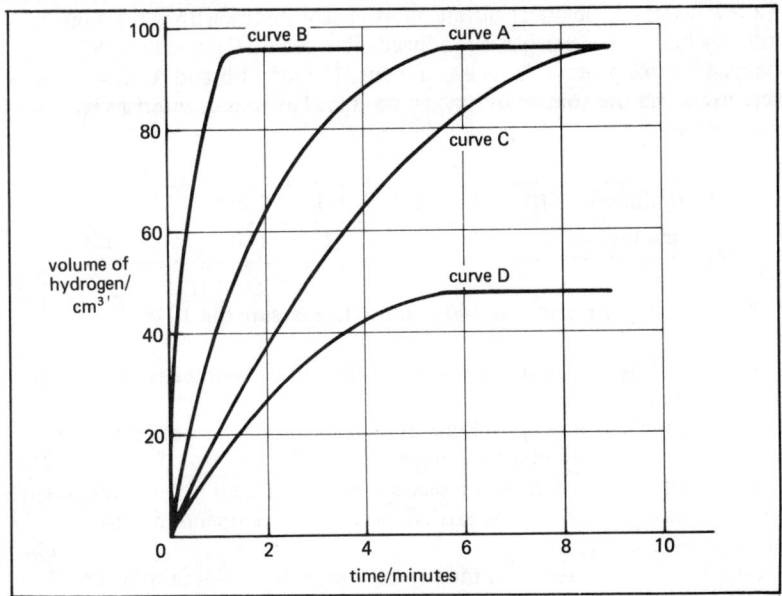

a Draw a labelled diagram of the apparatus you would use to carry out this experiment. (2)

b (i) Referring to the results shown in curve A, during which one minute period was the reaction proceeding most rapidly?

(ii) Explain what causes the reaction to proceed most rapidly during this period. (2)

c On the same graph are shown curves B, C and D, which were obtained from the same reaction by slightly altering the conditions.

(i) Which of the curves B, C or D would be obtained by using the same mass of powdered zinc in place of granules? Explain why this change results in the observed change in the curve.

(ii) Which of the curves B, C or D would be obtained by using acid of only half the molarity (but still present in excess)? Explain why this change results in the observed change in the curve.

(iii) Which of the curves B, C or D would be obtained by using half the mass of granulated zinc? Explain why this change results in the observed change in the curve.

(iv) Which of the curves B, C or D would be obtained by using the original hydrochloric acid warmed to a temperature about $10°C$ higher. Explain why this change results in the observed change in the curve. (4)

d (i) How many moles of hydrogen molecules were obtained in the experiment represented by curve A?

(ii) What mass of zinc must have been used to produce curve A? (2)

41. Decomposition of Bleach

If a few drops of cobalt(II) nitrate solution are mixed with a solution of ordinary bleach, oxygen gas is produced. The cobalt(II) nitrate is NOT changed by this process. In an experiment, 10 cm³ of brand A bleach were used, and the volume of oxygen produced at various intervals is given below.

Time/minutes	0	0·5	1·0	1·5	2·0	2·5	3·0
Volume O_2/cm³	0	94	124	142	150	150	150

a What piece of apparatus would you use to measure the 10 cm³ of bleach? (1)

b As the cobalt(II) nitrate is not changed in this experiment, what can you say about it? (1)

c Plot a graph showing the volume of oxygen (on the y-axis) against time (on the x-axis). Clearly mark the graph 'brand A'. (2)

d How many moles of oxygen molecules were collected in this experiment? (Assume the experiment to be carried out at room temperature and atmospheric pressure.) (1)

e Suppose that you repeated the experiment, using a bleach solution of the same concentration but approximately 10°C warmer than the original one. Sketch on your graph the new curve you would expect to obtain. Mark it clearly 'warm **A**'. (2)

f Brand **B** is another bleach, similar to brand **A**, but the active ingredient is only half as concentrated as in brand **A**. Assuming that an experiment with brand **B** was carried out in exactly the same way, with the same volumes of solutions, as the original (low temperature) brand **A** experiment, sketch on your graph the curve you would expect to obtain Clearly mark it 'brand **B**'. (2)

g Cobalt(II) nitrate, manganese(II) sulphate and many iron salts all have a similar effect on bleach solution. What do these compounds have in common which might make them all effective in this way? (1)

42. The rate of reaction: Zinc and Sulphuric Acid

Some zinc granules were carefully weighed out and dropped into a flask containing excess sulphuric acid. The mouth of the flask was quickly plugged with loosely-packed cotton wool. The mass of the flask and contents was then determined at half-minute intervals, with the results shown in the table:

Time/minutes	0	0·5	1·0	1·5	2·0	2·5	3·0
Total mass/g	70·550	70·495	70·470	70·460	70·455	70·450	70·450

The reaction which takes place may be represented by the following equation:

$$Zn(s) + H_2SO_4(aq) \longrightarrow ZnSO_4(aq) + H_2(g)$$

a Why does the mass of the flask and contents decrease as the experiment proceeds? (1)

b What would be the purpose of the loose cotton wool plug? (1)

c How long after the start was the reaction complete? (1)

d When was the reaction taking place most rapidly? (1)

e (i) Calculate the mass of hydrogen, and hence the number of moles of hydrogen molecules which were produced in this experiment.

(ii) Calculate the number of moles of zinc, and hence the mass of zinc which must have been used in this experiment. (4)

f Suppose that you were to repeat the experiment using the same mass of powdered zinc in place of the granules, everything else being kept exactly the same.

(i) In the long run, would you still expect the same loss of mass? State 'yes' or 'no' and explain how you arrive at your answer.

(ii) Would you expect the new reaction to take place faster, more slowly or at the same speed as the original one? Again explain how you arrive at your answer. (2)

43. The Decomposition of Potassium Chlorate(V)

When potassium chlorate(V), $KClO_3$, is heated at $600°C$ it decomposes into potassium chloride and oxygen. The equation for this reaction is:

$$2KClO_3 \longrightarrow 2KCl + 3O_2$$

Copper(II) oxide is said to act as a catalyst for the reaction. Some data for these compounds are shown in the following table:

	KCl	KClO$_3$	CuO
melting point/°C	772	368	1326
boiling point/°C	1407	–	?
solubility in water	soluble	soluble	insoluble

a Why is there no boiling point quoted in the table for potassium chlorate(V)? (1)

b (i) In what physical state will the potassium chlorate(V) be when it starts to decompose?

(ii) In what physical state will the potassium chloride be when it forms in this reaction?

(iii) Describe what you would expect to SEE as this reaction proceeds. (3)

c A catalyst may be defined as a substance which alters the rate of a chemical reaction, without itself being used up in the process.

(i) Draw a labelled diagram of the apparatus you would use to measure the rate of decomposition of potassium chlorate(V) in this reaction.

(ii) How would you make use of the apparatus you have just drawn in order to show that the rate of reaction is in fact increased by the addition of copper(II) oxide?

(iii) If you had a mixture of potassium chloride, copper(II) oxide and, possibly, potassium chlorate(V), how would you obtain pure copper(II) oxide from the mixture?

(iv) How would you prove that copper(II) oxide is not in fact used up in this reaction? (4)

d What is the maximum volume of oxygen which could be obtained, (measured at room temperature and atmospheric pressure) from 12·25 g of potassium chlorate(V) by heating it? (2)

Section 8
Reversible Reactions

44. Equilibrium

If carbon dioxide under pressure is allowed to mix with water in a sealed container, a dynamic equilibrium is set up, according to the following equation:

$$CO_2 + 2H_2O \rightleftharpoons H_3O^+ + HCO_3^-$$

The pH of the solution produced in this way is about 6.

a (i) Explain the meaning of both 'equilibrium' and 'dynamic' in the above statement.

(ii) If a little radioactive carbon dioxide gas were added to the equilibrium mixture, where would you expect the radioactivity to end up: in the gas, in the solution, or in both? Justify your answer. (4)

b Why does the statement at the beginning specify a *sealed* container? (1)

c (i) Explain why the solution produced in this experiment has a pH of about 6.

(ii) If the solution produced in this experiment is poured into an open beaker and allowed to stand, bubbles of carbon dioxide may be seen to form slowly and then escape. What will happen to the pH of the solution? Explain how you arrive at your answer.

(iii) Explain *why* bubbles of carbon dioxide escape if the solution is left to stand in this way. (5)

45. A Reversible Reaction of Iron

$$Fe_3O_4(s) + 4H_2(g) \rightleftharpoons 3Fe(s) + 4H_2O(g)$$

a Explain what the term 'reversible' means when applied to the above reaction. (1)

b What mass of iron would be produced by the complete reduction of 2·9 g of the iron oxide? (1)

c (i) If 232 g of the oxide were completely reduced by hydrogen, what would be the volume of the steam formed (calculated at r.t.p.)?

(ii) If all this steam condensed to water at room temperature, what would be its volume? (3)

d Iron forms another oxide, Fe_2O_3, which is also reduced by hydrogen.

    (i) Give the name of this oxide.

    (ii) Write the equation for this reduction. (2)

e   (i) What is the colour of Fe_3O_4?

    (ii) What is the colour of Fe_2O_3? (1)

f Fe_2O_3 is the main ore from which iron is extracted. Write a balanced equation to illustrate the chief reaction in the blast furnace in which the oxide is reduced to iron. (2)

46. Two Equilibria

Ammonium chloride, NH_4Cl, decomposes reversibly when heated in a test tube. Ammonia, NH_3, and hydrogen chloride, HCl, are formed. Ammonium sulphate, $(NH_4)_2SO_4$, similarly decomposes to NH_3 and H_2SO_4. Equations for these two reactions are shown below:

$$NH_4Cl(s) \quad \rightleftharpoons \quad NH_3(g) \quad + \quad HCl(g)$$

$$(NH_4)_2SO_4(s) \quad \rightleftharpoons \quad 2NH_3(g) \quad + \quad H_2SO_4(l)$$

a When ammonium chloride is warmed in a test tube for some time, no residue remains at the bottom of the tube, but white crystals are observed on the upper part of the tube.

    (i) What are these crystals which form on the upper part of the tube?

    (ii) Explain how these crystals are formed, and why they form only on the upper part of the tube. (3)

b When ammonium sulphate is warmed gently in a test tube, a liquid eventually forms in the tube. This liquid does NOT solidify when the tube is cooled to room temperature, and NO white crystals form on the upper part of the tube.

    (i) What is this liquid which eventually remains in the test tube?

    (ii) Explain how this liquid forms, and why ammonium sulphate appears to behave differently from ammonium chloride. (3)

c Suppose that pieces of moist red and blue litmus papers were placed at the mouth of the tubes in each of the experiments described above. Predict what, if anything, you would expect to happen to EACH piece of litmus paper when

    (i) ammonium chloride is warmed in the test tube,

    (ii) ammonium sulphate is warmed in the test tube. (2)

d A convenient way of making ammonia gas in the laboratory involves warming ammonium chloride with the hydroxide of a group I or II metal. Explain carefully what effect the hydroxide ions, OH^-, will have on the equilibrium whose equation is given at the beginning. State why this effect results in the formation of large quantities of ammonia gas. (2)

47. An Organic Reversible Reaction

The reaction summarised by the following equation occurs when the substances on the left hand side of the equation are warmed together with concentrated sulphuric acid.

$$CH_3COOH + C_2H_5OH \rightleftharpoons CH_3COOC_2H_5 + H_2O$$
$$\text{A} \qquad \text{B} \qquad \qquad \text{C}$$

a Name the substances, **A**, **B**, and **C**, which are shown in the equation. (2)
b (i) Explain what is meant by the $\rightleftharpoons$ sign in the equation.
(ii) Write an equation for any other reaction for which this sign is appropriate. (2)
c How would you recognise the formation of the organic product of the reaction? (1)
d What is the purpose of the concentrated sulphuric acid in the reaction mixture? (1)
e It is possible to make one of the starting materials by oxidation of the other.
(i) Copy and complete this statement:
'...... can be made by oxidising ....'
(ii) Name a suitable oxidising agent. (2)
f (i) What is the name of the group of compounds to which the compound represented by the formula $CH_3COOC_2H_5$ belongs?
(ii) Name a synthetic polymer which has similar structural units. (2)

48. An Industrial Reaction

Nickel is usually obtained industrially from impure nickel ores by heating the ore with carbon monoxide, a very good reducing agent. Combination then occurs between the nickel formed and excess gas to form a new compound called tetracarbonylnickel(O), the impurities not being affected. If this nickel compound is then heated to a higher temperature, the action is reversed and pure nickel is obtained. The equation for this reversible reaction is

$$Ni(s) + 4CO(g) \underset{463 \text{ K}}{\overset{364 \text{ K}}{\rightleftharpoons}} Ni(CO)_4(g) \text{ (pressure is atmospheric)}$$

a (i) What is the Celsius temperature for the forward reaction?
(ii) What is the Celsius temperature for the reverse reaction? (1)
b If one mole of molecules of any gas at s.t.p. occupies $22 \cdot 41 \text{ dm}^3$, what would its volume be at 364 K and the same pressure? Give the answer to the nearest whole number (notice that 364 and 273 have a common factor). (2)

c How many moles of carbon monoxide react with one mole of nickel atoms? (1)

In a certain experiment, 2000 kg of crude nickel ore were taken and heated with carbon monoxide at 364 K. After reduction was complete, a further $1 \cdot 2 \times 10^5$ dm^3 of carbon dioxide (measured at 364 K and atmospheric pressure) were required to convert the metal into tetracarbonylnickel(O).

d (i) How many moles of carbon monoxide were present in this volume of gas under the conditions mentioned?

(ii) How many moles of nickel atoms would combine with this volume of carbon monoxide?

(iii) What mass of nickel (relative atomic mass = 59) was present in the ore?

(iv) Calculate the percentage of nickel in the crude ore. (4)

e Carbon monoxide forms similar compounds with both iron and cobalt.

(i) What do the metals nickel, iron and cobalt have in common?

(ii) Whereabouts do these three metals occur in the periodic table? (2)

Section 9
Acids, Bases and Salts

---

49. Acids and Ions

Questions (**a**) to (**d**) refer to the following formulae for a number of molecules and ions:

$$SO_2, \; CO_2, \; H_2SO_4, \; HCl, \; SO_4^{2-}, Cl^-, H_3O^+ \text{ (or } H^+ \text{ (aq))}$$

a (i) What type of heat change occurs when sulphuric acid is mixed with water, exothermic or endothermic?

(ii) Describe how you would carefully prepare a dilute solution of sulphuric acid from water and concentrated sulphuric acid. (2)

b With two exceptions, all the molecules and ions shown above have the same effect on moist neutral litmus paper.

(i) What effect would they have?

(ii) Which are the two exceptions?

(iii) What particle, present in aqueous solutions, causes this similarity of behaviour?

(iv) By means of a suitable equation, show, for ONE of the relevant molecules/ions, how this particle is formed in aqueous solution. (4)

c (i) Give the formulae of all the molecules/ions above which could be identified by the addition of aqueous silver nitrate acidified with dilute nitric acid.

(ii) Write a suitable equation for the reaction which takes place in ONE case. (2)

d Give the formulae of all the molecules/ions above which could be identified by the addition of EITHER aqueous barium chloride acidified with dilute hydrochloric acid OR aqueous barium nitrate acidified with dilute nitric acid.

(ii) Write a suitable equation for the reaction which takes place in ONE case. (2)

50. Molarity and Titrations

A class was given an aqueous solution of hydrogen chloride (hydrochloric acid) and asked to determine its concentrations in g dm^{-3}. They were told that it contained approximately 9 grams HCl dm^{-3}. They were also provided with some anhydrous sodium carbonate which they knew would react with the acid according to the equation

$$Na_2CO_3 + 2HCl = 2NaCl + H_2O + CO_2.$$

a The first thing one student did was to calculate the approximate molarity of the acid.

(i) What is meant by molarity when referring to a solution?

(ii) Which of the following figures gives the approximate molarity of the acid: 0·2, 0·025, 0·05, 0·25, 0·5? (2)

b The student decided to make up 250 cm^3 of a 0·1 M solution of the sodium carbonate in water.

(i) Why do you think that this molarity was chosen?

(ii) How many grams of the anhydrous sodium carbonate should be weighed out? (2)

c He pipetted 25 cm^3 of the aqueous sodium carbonate into a flask and added a few drops of methyl orange which turned yellow. The acid was run in from a burette until the colour just became red, 20·4 cm^3 being needed. He repeated the titration more carefully several times and the volumes of acid used were 20·6 cm^3, 19.9 cm^3, 20·1 cm^3 and 20·0 cm^3 Some of these readings were rejected and an average value calculated from the rest.

(i) If the average calculated volume was 20·0 cm^3, which values were rejected?

(ii) How many moles of sodium carbonate are there in 25 cm^3 of a 0·1 M solution? (Leave your answer as a fraction if you wish).

(iii) How many moles of the acid would be neutralised by this number of moles of sodium carbonate (look at the equation)?

(iv) What was the volume of the acid which contained this number of moles?

(v) How many moles of the acid were present in 1 dm³ of the solution?

(vi) What was the exact concentration of the acid in g dm⁻³? (6)

51. An Electrical look at Neutralisation

25 cm³ of 0·05 M sulphuric acid (H_2SO_4) were put into a beaker and two platinum electrodes dipped into it. The electrodes were connected to an a.c. (alternating current) source of electricity and a suitable ammeter. Barium hydroxide solution, $Ba(OH)_2$, was added to the acid a little at a time with thorough stirring, the current being noted after each addition. In all, a considerable excess of barium hydroxide was added. The results are shown on the graph.

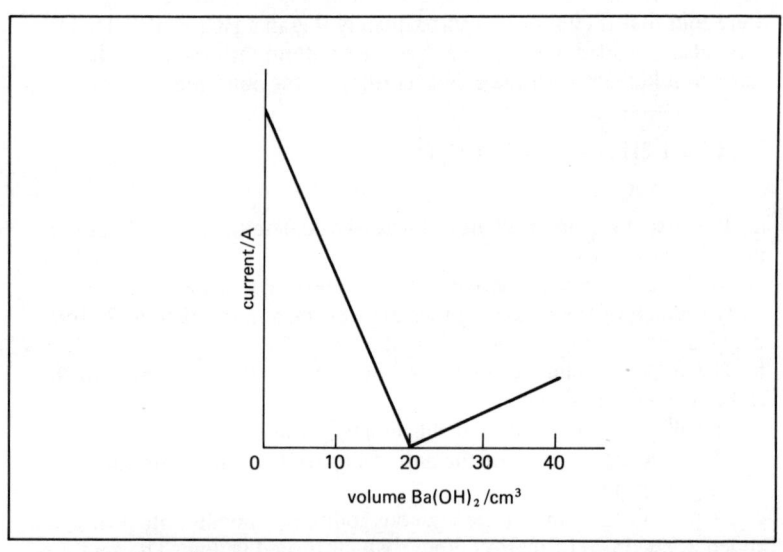

a What advantage was gained by using a.c. rather than d.c. (direct current) in this experiment? (1)

b As soon as some barium hydroxide was added to the acid, a white precipitate was formed.

(i) Give the name and formula of this white precipitate.

(ii) Assuming that barium hydroxide and sulphuric acid react in a 1:1 mole ratio, write a full equation for the reaction which takes place. (2)

c Give the name and formula of those particles which are present in aqueous solutions of all acids. (1)

d Suppose that some universal indicator had been added to the sulphuric acid before the start of the experiment. State roughly what pH you would expect it to show when the following volumes of barium hydroxide solution had been added: (i) 0 cm^3 (ii) 20 cm^3 (iii) 40 cm^3 (1)

e Ions must be present for a solution to conduct electricity. Explain TWO ways in which ions are being removed as the volume of barium hydroxide added increases from 0 to 20 cm^3. In each case write a suitable *ionic* equation. (4)

f Why does the conductivity of the solution increase when more than 20 cm^3 of barium hydroxide have been added? (1)

52. Equations and Molarity

50 cm^3 portions of a 0·5 M solution of lead nitrate, $Pb(NO_3)_2$, were added separately to six samples of a 0·5 M solution of a soluble sulphate. The masses of lead sulphate (relative molecular mass = 302·8) precipitated in each case were:

	A	B	C	D	E	F
Volume of sulphate solution cm^3	48·5	49·0	49·5	50·0	50·5	51·0
Mass of precipitate g		7·34	7·42	7·50	7·57	7·57	7·57

a (i) What fraction of a mole of lead sulphate had been formed when precipitation was complete? (Call the fraction $1/x$ and find the value of x)

(ii) How many moles of the sulphate solution were just sufficient to precipitate all the lead sulphate? (2)

b If the sulphate solution contained $5·5 \text{ g}/100 \text{ cm}^3$, find the mass of one mole of the compound. (1)

c The metal **Y**, the sulphate of which was used, belonged to group I of the periodic table. What is its
   (i) valency,
   (ii) relative atomic mass,
   (iii) name? (3)

d (i) Write a balanced full equation for the precipitation reaction, using the correct symbol for **Y** and the correct formula for its sulphate.

(ii) Rewrite this equation showing all the ions and their charges. (2)

e What colour would be shown by a platinum wire dipped in the sulphate solution and then placed in a Bunsen flame? (1)

f Write a balanced equation for the reaction of **Y** with water. Use the correct symbol for **Y**. (1)

53. Methods of Preparing Salts

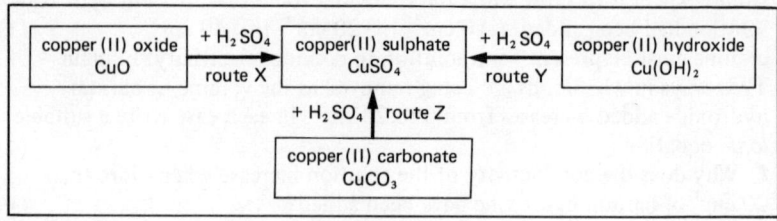

The above diagram shows three possible routes, X, Y and Z, by which copper(II) sulphate could be made from different starting materials by adding sulphuric acid solution. The following table gives some information about the solubility in water of a number of compounds of copper, sodium and lead.

Metal	Oxide	Hydroxide	Carbonate	Sulphate	Chloride	Nitrate
copper	insoluble	insoluble	insoluble	soluble	soluble	soluble
sodium	soluble	soluble	soluble	soluble	soluble	soluble
lead	insoluble	insoluble	insoluble	insoluble	insoluble	soluble

Use the above information in answering the questions which follow.

a In route Z a gas is formed as a by-product. What gas is this? (1)

b In an attempt to prepare *pure* crystals of copper(II) sulphate by route X, *excess* copper(II) oxide was added to a limited quantity of sulphuric acid and the mixture warmed.

(i) How would you be able to tell when excess oxide had been added?

(ii) How may the excess copper(II) oxide be removed?

(iii) Why is it necessary to add excess oxide in order to ensure that the product is pure?

(iv) Having removed the excess oxide, describe how you would then attempt to prepare pure hydrated crystals of copper(II) sulphate. (4)

c Route X (i.e. starting from the oxide) would not be so suitable if you wanted to make pure sodium sulphate instead of copper(II) sulphate.

(i) Explain why the route X is less suitable for making sodium sulphate than copper(II) sulphate.

(ii) Would routes Y or Z be any better for preparing sodium sulphate? Explain how you arrive at your answer. (2)

d A route different from routes X, Y and Z is often used to make lead(II) chloride. In this method, solutions of lead(II) nitrate and sodium chloride are simply mixed together and the precipitate of lead(II) chloride is filtered off.

(i) What is the other product of this reaction?

(ii) Why would this method be unsuitable for preparing samples of sodium and copper(II) chlorides?

(iii) Explain why routes X, Y and Z (using hydrochloric acid) would be less suitable for making pure lead (II) chloride than the method described above.   (3)

54. Two Sulphates of Sodium

An aqueous solution of sodium hydroxide, containing 8 g dm^{-3} was made up. A 25 cm^3 portion was placed in a flask and dilute sulphuric acid run into it until neutralisation was complete; 20 cm^3 were needed.

a What is the molarity of the sodium hydroxide solution?   (1)

b (i) Write the full equation for the neutralisation of sodium hydroxide solution by one mole of sulphuric acid.

(ii) How many moles of sodium hydroxide were used in the titration?

(iii) How many moles of sulphuric acid were needed for complete neutralisation?

(iv) What mass of sodium sulphate would have been obtained by evaporation of the solution to dryness?   (4)

The solution was not, in fact, evaporated, but a further 20 cm^3 of the sulphuric acid were added with stirring and the the solution was evaporated gently to form anhydrous sodium hydrogensulphate.

c (i) What is the molarity of the sulphuric acid?

(ii) Write the equation which represents the conversion of sodium sulphate to sodium hydrogensulphate by the method outlined above.

(iii) What mass of sodium hydrogensulphate was obtained?   (3)

d State ONE way in which a solution of sodium sulphate could easily be distinguished from one of sodium hydrogensulphate.   (1)

e State briefly (without using equations) TWO reactions which are common to both solutions.   (1)

55. Reactions of Solutions

Questions (a) to (f) refer to the following list of dilute aqueous solutions. In answering *choose from these solutions only* and give the formula in each case:

H_2SO_4	$Pb(NO_3)_2$	$CuCl_2$	$FeSO_4$
$Fe(NO_3)_3$	NH_3	HNO_3	Na_2CO_3

a Choose one solution which gives a precipitate with aqueous silver nitrate. Name this precipitate.   (1)

b Which two solutions give the same precipitate with aqueous barium chloride? Name this precipitate.   (1)

c Choose one solution which has a pH value greater than 7. (1)
d Choose three of the solutions which give precipitates with aqueous sodium hydroxide. State the colour of each precipitate. (3)
e Choose three of the solutions which react with metallic zinc. For each reaction, name ONE product. (3)
f Choose two solutions which react with each other. Name the products of this reaction. (1)

56. Analysis of a Salt

A white salt **A** dissolves in water to give a solution **B** which turns neutral litmus paper pink. When **A** is heated in a test tube, it disappears and gives off clouds of white fumes at the mouth of the tube. After cooling, a white deposit **C** is noticed near the mouth of the tube. A little of the solid **A** is placed in another tube with a pellet of sodium hydroxide and just covered with water. The tube is heated and a piece of filter paper, moistened with a little copper sulphate solution, which is held near the mouth of the tube goes a deep blue colour as a gas **D** is given off.

a (i) What is the name of the gas **D** which is evolved in the last reaction?
(ii) How would its solution affect neutral litmus paper? (2)
b Give the name and correct formula of the ion shown to be present in **A** by the reaction in which **D** is formed. (1)
c What is the name given to the type of reaction which occurs when **A** is heated? (1)
The gas **D** is passed over heated copper oxide and another gas **E** is formed.
d (i) Give the name of all the products formed in this reaction.
(ii) Write the balanced equation for this reaction.
(iii) How does the volume of **E** compare with that of **D**, both being measured under the same conditions? (3)
e (i) If a solution of **D** were added to aqueous iron(III) chloride, what would you notice?
(ii) Write an ionic equation for this reaction. (2)
Some of solution **B** is added to aqueous silver nitrate acidified with nitric acid and a white precipitate is formed. On standing, this precipitate becomes darker in colour.
f What ion is shown to be present in **A** by this result? (1)

57. Heat and Electricity on Salts

The nitrate **A** of a metal low in the electrochemical (reactivity) series was dissolved in water and added to aqueous sodium carbonate. A white precipitate **B** was formed in a colourless solution. When this mixture was filtered, and the filtrate gently heated and left to cool afterwards, white crystals **C** separated out.

When **B** was strongly heated, carbon dioxide was evolved and an orange-yellow solid was left which fused into the glass tube in which the reaction was being carried out. The same solid was formed on heating A alone, but brown fumes were then also observed.

Aqueous sodium bromide was added to a solution of **A**, and the heavy whitish precipitate **D** which was formed was dried and then placed in a crucible containing two wires connected to a battery. The crucible was heated and an attempt made to pass a current through **D**. When it had finally melted, a brown vapour came off around one electrode and a greyish metal **E** was formed at the other. This metal marked paper.

a   (i) Give the name and formula of **A**.
   (ii) Give the name of the TWO gases evolved on heating **A**.   (2)
b Write a balanced equation for the action of heat on either **B** *or* **C**.   (2)
c Give the name and formula of the gas which is given off in both cases by heating either **A** *or* **C**.   (1)
d Give the name and formula of the precipitate **D**.   (1)
e   (i) Give the name of the brown vapour which evolved in the last experiment
   (ii) State at which electrode this vapour came off.   (2)
f   (i) Give the name of the metal **E**.
   (ii) Why did **D** not conduct electricity until it was molten?   (2)

58. Aminosulphonic Acid

These questions concern a white crystalline solid known as amino-sulphonic acid, which has the formula H_3NSO_3. Two solutions of this substance were prepared, one in water and one in propanone (acetone).

a A pair of electrodes was dipped first into the propanone solution and then into the aqueous solution. It was found that whilst the propanone solution did not conduct, the aqueous one did.
   (i) What type of particles must have been present in the propanone solution?
   (ii) What type of particles must have been present in the aqueous solution?   (2)
b Pieces of dry universal indicator paper were dipped separately into the propanone and aqueous solutions. With the aqueous solution the paper turned red indicating pH = 1. The propanone solution indicated pH = 7.
   (i) What particles must be present in the aqueous solution?
   (ii) What conclusion can be drawn about the particles present in the propanone solution?   (2)
c The different results obtained with the two solutions suggest that the aminosulphonic acid reacts with one of the solvents. State which solvent this is, and explain carefully what happens in the reaction.   (2)

d Predict and explain what you would expect to happen if a piece of magnesium ribbon were dropped into
  (i) the aqueous solution,
  (ii) the propanone solution. (2)
e Predict and explain what you would expect to happen if small pieces of calcium carbonate were dropped into
  (i) the aqueous solution,
  (ii) the propanone solution. (2)

Section 10
Periodicity

59 and 60. The Periodic Table

The diagram below represents part of the beginning of the periodic table from element of atomic number 3 to element D of atomic number 20. The letters are **NOT** the actual symbols of the elements which occur in the positions shown, but they may be considered so for the purpose of these two items. Study the positions of the lettered elements and then answer items 59 and 60, *using only these letters in your answers.*

I	II	III	IV	V	VI	VII	O
			R	E		J	L
A		Q			G	M	
	D						

59.

a (i) How many electrons are there in the outer shell of an atom of **J**?
  (ii) How many electrons are there in the outer shell of an atom of **D**? (2)
b (i) What is the total number of electrons in an atom of **G**?
  (ii) What is the total number of electrons in an atom of **L**? (2)
c (i) Which two elements show a valency of three?
  (ii) Which of these two elements is the more likely to form salts containing a 3+ ion? (2)

d (i) How many neutrons are there in each atom of the common isotope of **R**?

(ii) How many neutrons are there in each atom of the common isotope of **A**? (2)

e (i) Give the letters of all the elements which are gases.

(ii) Give the letter of the gas which is the least reactive chemically. (2)

60. Remember to use only the lettered elements in the table in your answers.

a Which element exhibits three DIFFERENT valencies in the compounds it forms? (1)

b How many elements listed are metals? (1)

c Which element is capable of joining with itself to form long chains? (1)

d Which of the elements listed is/are polymorphic (allotropic)? (1)

e Which element listed forms a puckered ring molecule of eight atoms? (1)

f Which of the listed elements is the most electropositive (basic)? (1)

g Which is the element whose oxide is present in bauxite? (1)

h Give the letters of the two elements which you think would form the most strongly ionic salts. (1)

j (i) What would be the formula of the compound formed by **E** and **M**?

(ii) What type of structure would you expect this compound to have? (2)

61. The Periodic Table and the Behaviour of Elements

The symbols for a few elements have been marked in the outline below of the extended form of the periodic table. You are NOT expected to have a detailed knowledge of the chemistry of these elements, but you should have a general idea of their behaviour from their position in the periodic table. In answering the questions which follow, confine your answers to the elements whose symbols are shown above: you may use the symbols in your answers rather than the names of the elements.

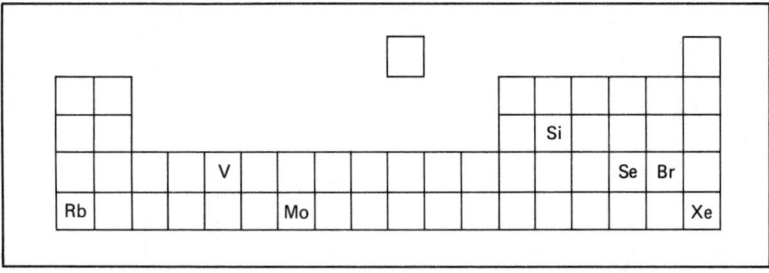

a Which element(s) would you expect to be gaseous at room temperature and pressure? (1)

b Which element(s) would you expect to conduct electricity very well?(1)
c Which TWO elements would you expect to be the most dense? (1)
d Which element(s) would you expect to have an alkaline oxide? (1)
e Which element(s) would you expect to form coloured compounds? (1)
f Which is the least reactive element marked? (1)
g Which of the marked elements would you expect to react most violently with sodium metal? (1)
h Which of the marked elements would you expect to react most violently with chlorine? (1)
j Francium, Fr, is in the same group as rubidium, Rb, but has a greater relative atomic mass. Would you expect Fr to be more or less reactive than Rb? Justify your answer. (1)
k Astatine, At, is in the same group as bromine, Br, but has a greater relative atomic mass. Would you expect At to be more or less reactive than Br? Justify your answer. (1)

Section 11
Metals and their Compounds

---

62. and 63. Sodium Chloride and its Reactions

The diagram below is to be used for items 62 and 63. It shows some of the compounds which can be obtained from sodium chloride; some of them are lettered for ease of reference and ⟶ means 'is converted into' or 'can form'.

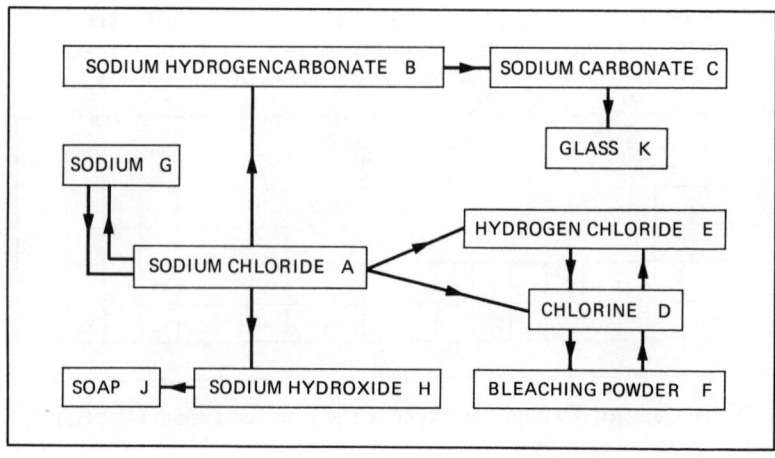

62.
a What is the name given to the crystalline form in which **A** is obtained
by mining? (1)
b What type of energy is used industrially in converting **A** to **H**? (1)
c Write equations for the direct conversion of
  (i) **H**(aq) to **B**(aq)
  (ii) **B**(aq) to **C**(aq) (2)
d Give the common name (or the chemical formula) for a crystalline
form of **C** used in most homes. (1)
e Give the name of any ONE naturally-occurring substance used with
C in forming glass (**K**). (1)
f (i) Give the name of a chemical which, on adding it to **A**, will
produce **E**.
  (ii) Give the name of the solid product formed in this reaction.
  (iii) **D** can be obtained directly from **A** by a chemical method which
involves heat. Give the names of the TWO substances which must be
added to **A** and heated with it to bring about this change. (4)

63.
a If **D** is passed into a test-tube of water a solution is obtained which
contains **E**; if a full tube of this solution is left for a while, corked,
bubbles of gas eventually appear.
  (i) Give the name of this gas.
  (ii) If a piece of neutral litmus paper is placed in a fresh solution
of **D** and left for two or three minutes, what will its colour then be?
  (iii) Write a balanced equation for the *complete* reaction between
D and water. (3)
b How is **E** produced from **D** industrially? (1)
c **D** can be obtained from a concentrated solution of **A** without the
use of any other chemicals.
  (i) Name the method that could be used.
  (ii) What other products would be either liberated or formed in
solution? (3)
d When converting **H** to **J** in the laboratory, **H** is dissolved in ethanol
and heated with a suitable organic compound. Give an example of an
organic compound which could be used. (1)
e **D** can be obtained from **F** by the addition of a dilute mineral acid.
Assuming the formula, $CaOCl_2$, for **E**, write a balanced equation for the
reaction which takes place when an acid (choose any you wish) is added
to it. (1)
f **A** can be synthesised from **G** + **D**; is this process exothermic or
endothermic? (1)

64. Calcium Compounds.

Slaked lime is a caustic alkali which is only slightly soluble in water. Mixed with sand and water it is used in large quantities by builders for making mortar and this substance, when exposed to the air, gradually hardens. The equation for this reaction is

$$Ca(OH)_2(s) + CO_2(g) \longrightarrow CaCO_3(s) + H_2O(l)$$

a What is the common name given to a solution of calcium hydroxide and for what purpose is it used in the laboratory? (1)
b (i) Give the name of the substance which causes mortar to harden.
(ii) What type of chemical reaction is this? (Look at the equation). (2)
c Why do the walls of newly-built houses often appear damp for some time even in dry weather? (1)

Slaked lime is produced by adding water to calcium oxide (lime), the mixture becoming hot when the water is first added. Lime itself is used in the preparation of calcium carbide (CaC_2) by heating it with coke in an electric furnace.

d In view of the temperature change occurring, how would you describe in one word the slaking of lime? (1)
e (i) Suggest ONE reason why an electric furnace is used in the preparation of calcium carbide rather than direct heating.
(ii) In this reaction the coke is oxidised to carbon monoxide; write a balanced equation to illustrate the complete reaction in the furnace. (2)

Calcium carbide reacts with water to form calcium hydroxide and an industrial gas called ethyne (a hydrocarbon) which are the only products.
f Write the equation for this reaction and hence suggest the formula for ethyne. (1)
g Now write a balanced equation for the burning of ethyne in air and name the products formed. (2)

65. Heating Copper Carbonate

0·31 g of pure copper(II) carbonate was heated in a test tube at about 600°C until decomposition was complete. The volume of gas evolved (measured at room temperature and pressure) was 60 cm^3. The gas was passed into a measuring tube, which was then stood in a dish of aqueous sodium hydroxide.

a (i) Explain why the volume of gas in the measuring tube decreased on standing.
(ii) Write a suitable balanced equation to represent what was happening in the measuring tube. (2)
b A black powder was left in the test tube after heating the copper(II) carbonate; calculate the mass of this solid. (2)

54

c (i) What was the mass of the gas given off?

(ii) Calculate the density of this gas in g cm^{-3} (3)

Ammonia gas was next passed into the test tube containing the powder test which was heated and the solid changed to a powder of a different colour.

d (i) What was the name of this new powder?

(ii) What was its mass? (2)

This powder was then heated in a stream of oxygen until the reaction was complete. (2)

e By how much had the powder increased in mass when the reaction was complete? (1)

66. The Economics of Corrosion

'Steel pipes or framework and the hulls of ships continually in contact with water may be connected to blocks of magnesium. These blocks are cheaper to replace than it is to have to effect repairs to the iron- or steel-work or to the bronze propellors of a ship. . . . Copper or brass bolts cannot be used with iron joists or plates exposed to water. . . . The steel or iron to be protected can be made the negative pole of an electric circuit which includes a large direct current generator. The anode may be scrap iron and this slowly dissolves.' (From *Metals* by O.J. Simpson.)

a (i) What other metal is often used instead of magnesium?

(ii) Where do these two metals occur in the electrochemical series relative to iron? (2)

b Why do you think it is cheaper to replace one of these blocks (which are quite large) than it is to repair or replace a rusty iron plate, when the cost of iron may be very much less than that of magnesium? (1)

c What is likely to happen to a copper-bottomed boat working in an estuary if the plates are fastened with iron rivets? (1)

d Name the two main elements present in

(i) bronze,

(ii) brass,

(iii) steel. (3)

e Corrosion involves the movement of electrons, the corroded metal going into solution as an ion. Which metal in each of the following pairs *loses* electrons if the two metals are in contact and exposed to damp or wet conditions?

(i) magnesium/iron

(ii) copper/iron (2)

f Consider an electric circuit in which an iron structure is made to act as the cathode so that it may be protected from corrosion. Write a balanced ionic equation to show what happens to the anode if this is made of scrap iron. (1)

67. Iron and Chlorine

The apparatus shown below was used to prepare a sample of anhydrous iron(III) chloride, $FeCl_3$.

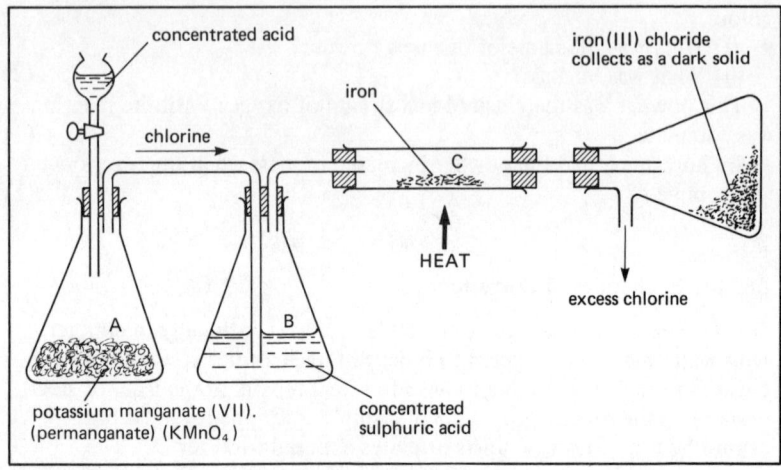

a (i) Name the acid which is run into flask A.

(ii) What is the purpose of the concentrated sulphuric acid in flask B? (1)

b Write an equation for the reaction taking place between iron and chlorine. (1)

c 0·5 mole of chlorine molecules, Cl_2, is passed over excess iron in tube **C**.

(i) What volume would this chlorine occupy at room temperature and pressure?

(ii) If this chlorine reacted completely with the iron, what mass of iron would be used up?

(iii) How many moles of iron(III) chloride would be formed? (3)

d (i) Why is it dangerous to allow any excess chlorine left over from the experiment to escape into the room?

(ii) How would you carry out the experiment to avoid this danger?

(iii) How would you detect whether the excess chlorine was escaping? (3)

e You have been provided with a dark solid which might be iron(III) chloride.

(i) Describe one test you would carry out to decide whether the solid was an iron(III) compound.

(ii) Describe one test you would carry out to decide whether the solid was a chloride. (2)

68. Cobalt, a Transition Metal

This question is concerned with the transition metal, cobalt, the chemistry of which may not be familiar to you. That is not important as the diagram below summarises a series of reactions which may be carried out with cobalt and its compounds.

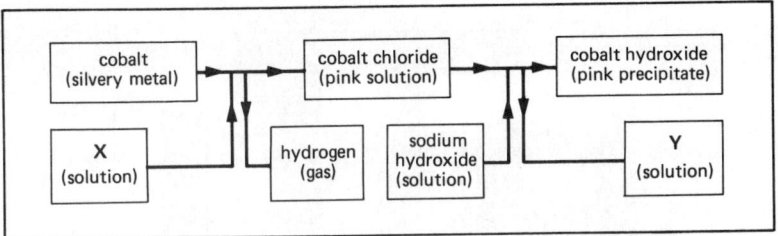

a (i) Predict TWO ways in which you would expect cobalt to differ in its properties from sodium.
(ii) Predict ONE way in which you would expect cobalt to resemble sodium in its properties. (3)
b (i) Identify the substance **X** in the above diagram.
(ii) Identify the product **Y** in the above diagram. (2)
c What conclusion can you draw about the position of cobalt in the electrochemical (reactivity) series from the information in the above diagram? (1)
d State TWO ways, typical of a transition metal hydroxide, in which cobalt hydroxide differs from sodium hydroxide. (2)
e If a piece of filter paper is soaked in the cobalt chloride solution and allowed to dry, the colour turns from pale pink to blue. What use is made of this in chemistry? (1)
f Predict the appearance of cobalt carbonate, giving a reason for your prediction. (1)

69. The Production of Aluminium

Industrially, aluminium is made from its oxide by dissolving the latter in molten cryolite and electrolysing the mixture at a temperature of about 1000°C, using a current of some 30 000 A. Both electrodes are made of carbon, although the anode is gradually consumed during the electrolysis and has to be continually replenished. The cryolite behaves merely as a solvent, and the reaction which takes place may be written as

$$2Al_2O_3 \longrightarrow 4Al + 3O_2$$

A diagrammatic representation of the cell used is shown below:

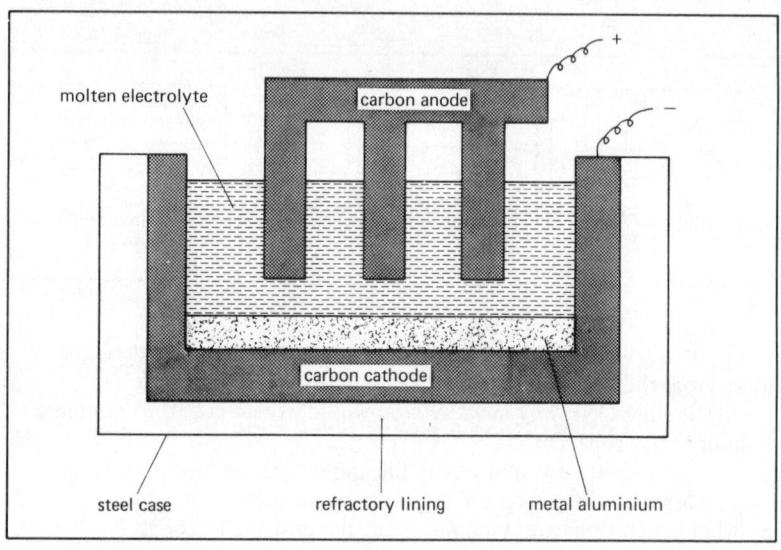

This table gives some information about aluminium and its compounds. Use this information in answering the questions which follow:

Name	Formula	Melting point/$^\circ$C	Boiling point/$^\circ$C
aluminium	Al	659	2447
aluminium chloride	$AlCl_3$	sublimes at 180	
aluminium oxide	Al_2O_3	2040	2980
cryolite	Na_3AlF_6	1000	?

a What is the purpose of the refractory lining? What properties must a suitable material have? (1)
b Sodium and magnesium are obtained by the electrolysis of molten sodium and magnesium chlorides. Explain why aluminium cannot be similarly produced from aluminium chloride. (1)
c Suggest why aluminium oxide is dissolved in molten cryolite, rather than being electrolysed in the pure state. (1)

58

d In what physical state will aluminium be formed in the electrolytic cell? (1)

e Suggest a reason why the anode gets consumed during electrolysis. (1)

f (i) How many faradays (moles of electrons) will be required to produce 1 mole of aluminium atoms from its oxide?

(ii) What quantity of electricity is consumed by the cell in 1 hour?

(iii) How many faradays are used by the cell in 1 hour?

(iv) How many moles of aluminium atoms will be produced by the cell in 1 hour?

(v) What mass of aluminium will be produced in 1 hour? (5)

70. Zinc

a The addition of sodium hydroxide solution to aqueous zinc sulphate produces a white precipitate which dissolves on adding excess of the alkali.

(i) Give the name and formula of the precipitate.

(ii) State the name of the substance formed when the precipitate dissolves.

(iii) Give an equation which represents the action of excess alkali on the precipitate. (3)

b The addition of aqueous sodium carbonate to a solution of zinc sulphate also produces a white precipitate **A**.

(i) Give the name of this precipitate.

(ii) Give an equation to show how it is formed. (2)

c If **A** is heated strongly it changes from white to yellow with loss of mass, but on cooling it again becomes white.

(i) What is the name of the solid formed which is yellow when hot and white when cold?

(ii) Write a balanced equation for the action of heat on **A**. (2)

d If 2·5 g of solid **A** is heated to constant mass, what is the mass of the substance formed? (1)

e (i) State one widespread use of zinc (not a laboratory use).

(ii) Give the reason why zinc is considered suitable for this use. (2)

71. Metals and Water

A metal **A** reacted with water fairly rapidly to evolve a gas **B**. This was burned at the end of a jet so that the flame touched the surface of a flask through which cold water was flowing. A liquid dripped from the surface into a conical flask containing a greyish-white powder which turned blue and eventually a solution was formed. The conical flask was then set up for distillation and, after heating, a colourless distillate **C** was obtained.

a (i) Give the name of the gas **B**.

(ii) Write a balanced equation for its burning in air. (2)

b (i) Is the distillate **C** a mixture or a compound?

(ii) What would you expect its boiling point to be? (2)

The solution left in the container in which the metal had reacted with water had become cloudy. A platinum wire dipped in this solution gave flashes of a brick-red colour when it was placed in a Bunsen flame. When the solution was evaporated, a white powder **D** was obtained and this glowed brightly when heated very strongly.

c (i) Which metal is identified by the brick-red colour in the flame?

(ii) What class of compound is formed by the reaction of this metal with water?

(iii) Would the pH of the solution formed be greater or less than seven?

(iv) What is the name of the white powder **D**? (4)

d (i) In which groups of the periodic table are found those metals which react with water?

(ii) What kind of structure (ionic or covalent) do the compounds which these metals form with water possess? (2)

72. Transition and other Metals

The chloride of a metal **A** was dissolved in water and added to a solution of the nitrate of a metal **B**, lower in the electrochemical series than the metal present in **A**. A white precipitate **C** was produced which did not dissolve in 2M nitric acid, but was soluble in aqueous ammonia: when it was exposed to the air it darkened in colour.

a Give the name and formula of EITHER precipitate **C** OR compound **B**. (1)

A solution of the nitrate of a different metal **D** was then added to a fresh portion of the original chloride solution and a heavy white precipitate **E** was formed as a suspension and, on filtering, this gave a coloured filtrate **F**. The addition of aqueous sodium hydroxide to **F** produced a pale blue precipitate **G** and this changed to a black colour on heating.

b (i) What metal is present in **A**?

(ii) Give an ionic equation for the reaction of **A** with **B** to show the formation of **C**. (2)

c From the information given

(i) identify the precipitate **E**,

(ii) suggest a suitable metal which might be present in **D**,

(iii) write an equation, ionic or molecular, showing the formation of E. (3)

d Is **F** likely to be an ionic compound or not? (1)

e (i) Write an equation representing the change which occurs when **G** is heated.

(ii) State the two changes you would SEE on gradually adding an excess of aqueous ammonia to **G**. **(3)**

Section 12
Non-metals and their Compounds

73. Hydrogen Chloride and its Solution

a Imagine that you are required to prepare some dry hydrogen chloride.
   (i) Name the starting materials that you would use.
   (ii) State whether heat would be necessary or not.
   (iii) Name a suitable drying agent for the hydrogen chloride. **(4)**
b The dry gas is bubbled into water to give solution **A** and into toluene (a hydrocarbon) to give solution **B**.
   Copy the following table and fill in the blanks to show the reactions, if any, between the substances named and solutions **A** and **B**. Equations are not required. If no reaction occurs, write 'none' in the appropriate box. Any products should be named.

Substance	Solution A	Solution B
metallic zinc		
metallic copper		
calcium carbonate		
dry pH paper or dry, neutral litmus paper		**(4)**

c How can you account for any difference in the behaviour of solution **A** and solution **B** in the table above? **(2)**

74. An Investigation of Salt Gas

According to a certain textbook, 'salt gas' may be made by dripping concentrated sulphuric acid onto lumps of rock salt. If the resulting gas is passed over hot iron, then crystals form in the combustion tube and hydrogen passes out at the end. The book recommends the following apparatus:

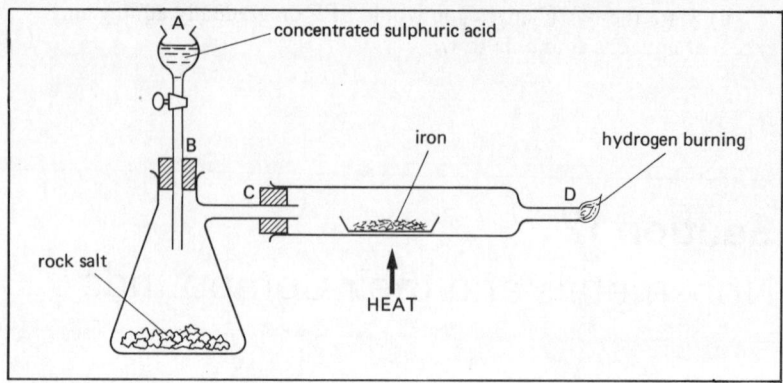

a What would be the advantage in allowing the concentrated sulphuric acid to drip slowly on to the rock salt, rather than adding it all in one go?

(1)

b The instructions for the experiment state that the gas should be allowed to pass through the apparatus for a few minutes before starting to heat the iron. What might happen if the iron in it was heated straight away? (1)

c As shown in the diagram, the book suggests that the hydrogen should be burned as it comes out of the combustion tube. What precaution(s) should you take before attempting to light the hydrogen, and why should this/these be necessary? (1)

d When the experiment was tried, it proved impossible to light the hydrogen, although there were signs of crystals forming in the tube. It was suggested that the salt gas was being generated too slowly. Suggest how it could be made more rapidly other than by dripping the acid on to the rock salt more quickly. (1)

e Increasing the speed of the salt gas production had little effect; the hydrogen still could not be lit. It was then thought that the hydrogen might have so much excess salt gas mixed with it that this was preventing it from burning. Salt gas may be trapped by passing it through granules of soda-lime. At which of the points A, B, C or D would you put a tube filled with soda-lime? (1)

f Having still failed to get the hydrogen to burn, it was decided to try to collect the gas in a suitable container before trying to test it. Draw a diagram of the apparatus you would attach at the point D in order to collect a sample of the gas suspected of being hydrogen. (1)

g How would you test the collected gas to prove that it is in fact hydrogen? (1)

h The crystals were thought to be iron chloride. In a repeat of the experiment, the following masses were determined:

mass of empty combustion tube   = 10·21 g
mass of combustion tube + iron   = 13·01 g
mass of combustion tube + iron chloride = 16·56 g

(i) How many moles of iron were used in the experiment? Show your working.

(ii) How many moles of chlorine were combined with this amount of iron? Again, show your working.

(iii) What is the formula of this iron chloride? (3)

75. Chlorine

Chlorine is a very active element which has a strong affinity for hydrogen and which can form many compounds by synthesis.

a (i) What is meant by 'affinity for hydrogen'?

(ii) What is meant by 'synthesis'? (1)

b What compounds are formed when chlorine is passed over

(i) heated iron,

(ii) heated sodium?

Give the names of the substances formed and state the valencies shown by the two metals in the compounds. (2)

c Dutch metal contains both copper and zinc and looks like the very thin sheets of gold leaf that some painters use in gilding. If a sheet of it is dropped into a jar of chlorine it catches fire and produces a smoke which setttles down as a greyish powder.

(i) What is the name given to a metallic substance which contains more than one metal?

(ii) Why do you think that sheets of Dutch metal take fire in chlorine whereas zinc and copper powder are not affected if dropped into chlorine?

(iii) What substances are likely to be present in the grey powder produced? (3)

d Turpentine is a liquid with the formula $C_{10}H_{16}$ and if it is warmed in a test tube and poured into a jar of chlorine a reddish flame and clouds of greyish smoke are produced.

(i) To what class of organic compounds does turpentine belong?

(ii) What substances are likely to be formed when the turpentine burns which would cause the greyish smoke?

(iii) How many moles of chlorine molecules would react completely with one mole of turpentine?

(iv) Is the turpentine oxidised or reduced when it reacts with the turpentine? Give a reason for your answer. (4)

76. Reactions of the Halogens

a Give the name and formula of the simple ion formed from

(i) a solid halogen,

(ii) a liquid halogen. (2)

b Fluorine is obtained by the electrolysis of fused potassium fluoride. Write the balanced equation to represent this reaction. (1)

c (i) Show by an ionic equation how a chlorine atom becomes an ion.

(ii) Is this change one of oxidation or reduction – and why? (2)

d (i) In which of the following situations will a reaction take place?
  (1) Cl_2 + KI(aq)
  (2) I_2 + KBr(aq)
  (ii) Write the balanced equation for any reaction which occurs.　(2)
e Which halogen
  (i) is present in the non-stick coating on kitchen pans,
  (ii) is used for killing bacteria in drinking water,
  (iii) is often added to table salt to overcome any deficiency in the body,
  (iv) forms a silver salt used in photography?　(2)
f Halogens react with aqueous alkalis. When chlorine reacts with a hot solution of sodium hydroxide it forms two compounds with the formulae $NaCl$ and $NaClO_3$, along with water. Write the balanced equation for this reaction.　(1)

77. Hydrides

Hydrides are compounds of hydrogen with one other element.

a Give the name of one hydride which dissolves in water for form a solution with a pH less than 7. Write a suitable equation for the reaction which takes place.　(2)
b Give the name of one hydride which dissolves in water to form a solution with a pH greater than 7. Write a suitable equation for the reaction.　(2)
c Give the name of one hydride which burns readily in air. Write a suitable equation for the reaction which takes place.　(2)
d Lithium/Li^+ hydride has an ionic structure $Na^+ H^-$. Draw a 'dot and cross' diagram to show how you would expect the electrons to be arranged in lithium hydride.　(1)
e (i) Give the name and formula of one hydride which is composed of molecules containing three or more atoms.
  (ii) Draw a 'dot and cross' diagram to show how the electrons are arranged in molecules of this hydride.
  (iii) Draw a simple diagram to indicate the approximate shape you would expect molecules of this hydride to have.　(3)

78. Hydrides, Familiar and Unfamiliar

A hydride is a compound containing hydrogen and one other element only.

a Name, in *each* of the following cases, one hydride which is a gas at room temperature and pressure and which will
  (i) dissolve readily in water to form an acid solution,
  (ii) dissolve readily in water to form an alkaline solution,
  (iii) decolorise bromine water.

(iv) In one case only, write a balanced equation for what is taking place. (2)

b Explain briefly, by naming the starting materials and stating the reaction conditions, how TWO of the hydrides selected in (a) can be prepared in the laboratory. (2)

c Sodium hydride, NaH, is formed as the only product when hydrogen is passed over heated sodium. It is a white crystalline solid which melts at $750°C$.

(i) Write a balanced equation for its formation.

(ii) Do you think that sodium hydride exists as molecules or as an ionic compound? Give reasons for your answer. (2)

d Carbon forms a very large number of hydrides which are grouped in several families, or homologous series. Name TWO such families of hydrides and give the name and formula of one member of each. (2)

e Two hydrides of phosphorus (relative atomic mass = 31) have relative molecular masses of 34 and 66 respectively.

(i) What are their chemical formulae?

(ii) If these are the first of two members of a homologous series what would be the formula of the third member of the series? (2)

79. Air: In Spaceships and on Earth

Two different methods have been used by American and Russian scientists to purify the air inside space ships. They are summarised by these equations:

American method: $\quad 2LiOH + CO_2 \longrightarrow Li_2CO_3 + H_2O$

Russian method:
$$2NaO_2 + CO_2 \longrightarrow Na_2CO_3 + 1\tfrac{1}{2}O_2$$
$$Na_2CO_3 + H_2O + CO_2 \longrightarrow 2NaHCO_3$$

Overall $\quad 2NaO_2 + H_2O + 2CO_2 \longrightarrow 2NaHCO_3 + 1\tfrac{1}{2}O_2$

a (i) What are the two main gases in normal air? Give their approximate percentages by volume.

(ii) Give the names of three other gases which are found in pure air. (2)

b (i) By what means is carbon dioxide naturally removed from the air on earth?

(ii) Why is it necessary to remove carbon dioxide from the air in space ships? (2)

c Both the American and Russian processes produce useful by-products. State what that by-product is, and why it may be useful in a space ship in

(i) the American process,

(ii) the Russian process (2)

d Calculate the mass of chemicals required to remove 1 mole of carbon dioxide from the air using
    (i) the American process,
    (ii) the Russian process. (3)
e Do you consider either process to be superior to the other? Give your reason(s). (1)

80. Different kinds of Oxides

The Venn diagram here shows how acidic and basic oxides fit into the general family of oxides.

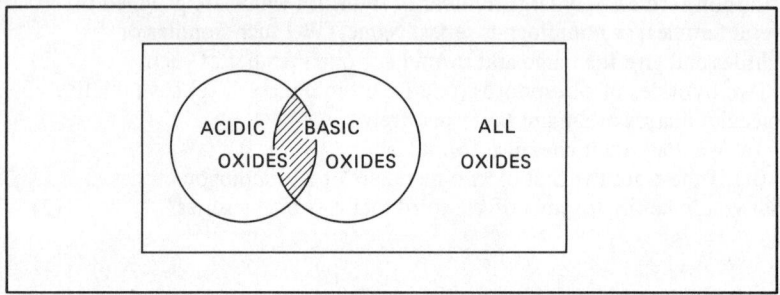

a Suppose you were asked to prepare an acidic oxide, but not to collect it.
    (i) Give the name of a suitable oxide.
    (ii) How would you prepare it? (1)
b What ONE change would you make if you had to prepare a basic oxide in a similar manner? (1)
c There are other methods by which these oxides can be made, starting from suitable compounds.
    (i) Give the names of THREE different classes of compounds from which you could obtain a basic oxide by heating.
    (ii) Give the name of ONE such compound with its formula.
    (iii) Write the balanced equation for the action of heat on a suitable compound from one of the OTHER classes you have mentioned in (i). (4)
d Acidic oxides can also be prepared by heating suitable compounds, but other methods are usually used.
    (i) Give one method for obtaining an acidic oxide in the laboratory which differs from your answer to question (a) and does not involve heating, naming the substances to be used.
    (ii) Write the balanced equation for this reaction. (2)
e Give the name of one compound, along with its formula, which, on heating, yields BOTH an acidic oxide and a basic oxide. (1)
f In the diagram the two circles overlap. EITHER state the name given to the type of oxide which could be placed in this overlap OR give the name or formula of any one oxide which could be correctly placed in this overlap. (1)

81. Water: The Essential Compound

a (i) Give the name of one element which will react with water to liberate hydrogen.

(ii) Write the balanced equation for this reaction. (2)

b Give the name of another, different method of obtaining hydrogen from water. (1)

c What do you think is the main reason why neither of the above methods is used generally for the production of hydrogen on a very large scale? (1)

d Photosynthesis is a process which involves water along with sunlight and chlorophyll.

(i) Which gas in the atmosphere is also involved in this reaction?

(ii) What gas is produced in the reaction?

(iii) What type of organic compound is produced in the reaction? (3)

e Liquid hydrogen chloride is a covalent compound which has no effect on dry neutral litmus paper. If water is added there is a change of colour in the paper.

(i) Give the name of the compound formed by the addition of water to liquid hydrogen chloride and state the type of structure it has.

(ii) Give the name and formula of the particle which actually causes the colour change in the litmus.

(iii) Write a simple balanced equation to show what has happened to the original hydrogen chloride molecules. (3)

82. Solubility

The solubility of a solid is usually given as a figure which represents the maximum number of grams of the solid which will dissolve in 100 g of water to form a saturated solution at a particular temperature. The solubility of a gas is often stated as the maximum number of volumes of the gas which will dissolve in one volume of water at atmospheric pressure at a stated temperature.

a (i) Give the name and formula of a chemical compound which is more soluble in hot water than in cold.

(ii) Give the name and formula of a chemical compound which is less soluble in hot water than in cold. (2)

b If the solubility of chlorine at room temperature and atmospheric pressure is 2·5, what mass of the gas will dissolve in 500 cm³ of water? (1 dm³ of chlorine under these conditions has a mass of 3·2 g.) (1)

c The solubility of a solid **X** at 80°C is 100 and at 20°C is 20. How could you tell by looking at it, whether a solution of **X** which had been standing at 20°C for some hours after stirring was really saturated or not? (1)

d (i) If you wished to make a solution of **X** which was saturated at 80°C, using 100 g of water, to approximately what temperature would you heat the water?

(ii) What mass of solid should be added to make sure that the resulting solution is saturated? (2)

e (i) What would be the total mass of a saturated solution of **X** at 80°C if 100 g of water were used?

(ii) How much water would be present in 40g of this saturated solution at 80°C?

(iii) How much of the solid **X** would be present in 40 g of this saturated solution at 80°?

(iv) How much of the solid **X** could be held in solution at 20°C by the same mass of water as in (ii) above? (3)

f If the 40 g of saturated solution at 80°C is now cooled down to 20°C, what mass of **X** will separate out? (1)

83. Solubility of Salts

The solubility of potassium nitrate in water may be measured as follows:

10g of potassium nitrate are transferred to a boiling tube and 6 cm^3 of water are added. The mixture is warmed until all the solid dissolves, and is then allowed to cool with constant stirring. The temperature at which crystals appear is noted. The mixture is then warmed again to redissolve the crystals. It is allowed to cool as before, and the crystallisation temperature is again noted. A further 2 cm^3 of water are added to the boiling tube and the new crystallisation temperature is again determined twice. Further portions of water are added and, each time, the crystallisation temperature is determined. The results are shown on the graph which plots the solubility in *moles* of solute per 100 g of solvent. The graph also shows the solubilities of potassium chloride, sodium chloride and sodium nitrate.

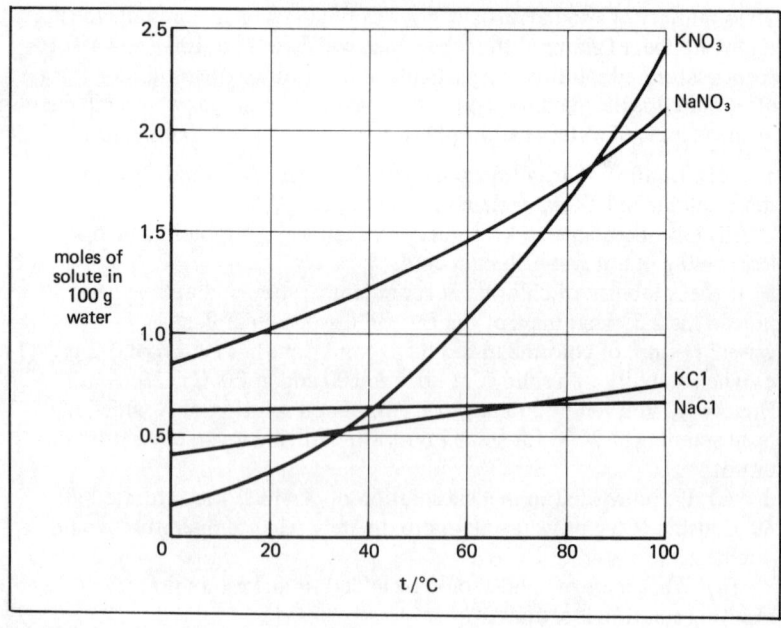

a (i) What piece of apparatus would you use for measuring out the water?

(ii) Why is constant stirring necessary when cooling the hot solution?

(iii) Why is the crystallisation temperature determined twice for each concentration? (3)

b At what temperature would a solution of 10 g of potassium nitrate, KNO_3, in 6 cm^3 of water crystallise? Show your working. (2)

c What is the maximum mass of sodium chloride, $NaCl$, which can be dissolved in 1 kg of water at 30°C? (1)

d If a solution containing 1·5 moles of sodium nitrate, $NaNO_3$, in 100 g of water is cooled from 80°C to 20°C, how many grams of solid will crystallise out? (2)

e If 0·5 mole of $NaNO_3$ is added to a solution of 0·5 mole of KCl (potassium chloride) in 100 g of water at 40°C nothing happens. When the solution is cooled to 20°C however, a white solid crystallises out. What is this solid? Explain why this happens. (2)

84. The Combustion of Carbon

The diagram below represents a coke fire burning and the reactions which occur are the same as those taking place at the bottom of a blast furnace for the production of iron.

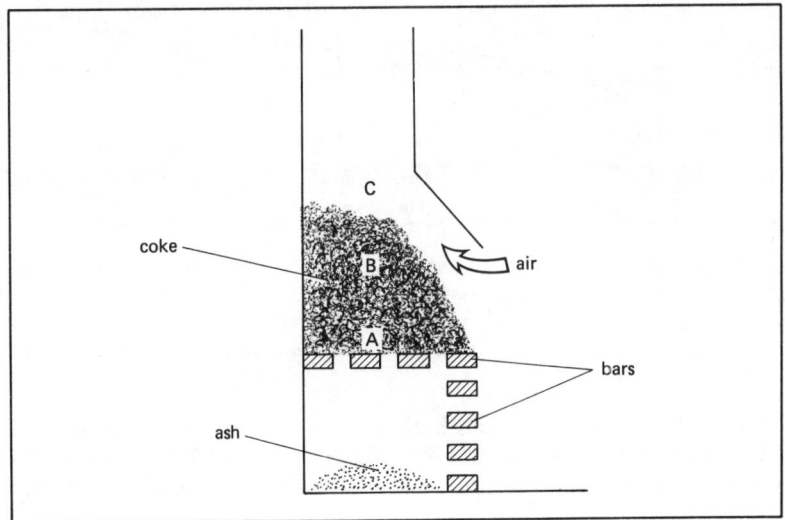

a What is the reason for having bars in the grate rather than a solid base? (1)

b Write an equation (balanced) for the reaction which occurs at the base of the fire, A, and name the gas **X** produced. (1)

c A further reaction occurs within the hot coke at B involving both oxidation and reduction.

(i) What is the chemical name of the substance which is oxidised in this reaction?

(ii) Write the balanced equation for the reaction taking place at B.　(2)

d (i) What colour (other than yellow) are you most likely to see in the flames at C?

(ii) What is the name of the gas **Y** formed at B which is now reacting at C?

(iii) Write the equation for the reaction at C.　(3)

e In the blast furnace this same gas **Y** brings about the production of iron from the ore put into the furnace. Does it do this by oxidation or by reduction?　(1)

f One of the other raw materials put into the blast furnace produces gas **X** on decomposition.

(i) Name this material and give its formula.

(ii) Give the main reason for the use of this substance in the extraction of iron.　(2)

85. Fuel Gases from Carbon

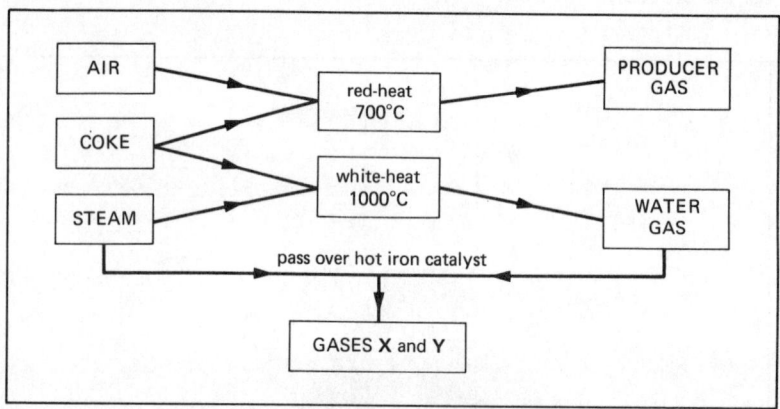

Producer gas and water gas are both made from coke, the former by blowing air through a deep bed of red-hot coke (about 700°C) and the latter by treating white-hot coke (about 1000°C) with steam.

a What are the main constituents (two in each case) present in
(i) producer gas,　(ii) water gas?　(2)

b One of the two reactions is exothermic and the other endothermic.

(i) Which reaction is endothermic?

(ii) If the heat change per mole of carbon is x kJ, write the equation

for the production of water gas, including the ΔH term with the correct
sign. (2)

c If equal masses of carbon are fully converted into the two gases named
above

(i) which gas would have the higher calorific (heating) value?

(ii) Explain how you arrive at your answer. (2)

d The two constituents of water gas can be separated by passing the
mixture with extra steam over a heated iron catalyst. One of the gases
is oxidised and can then be dissolved out under pressure in water (or
more easily in aqueous potassium hydroxide).

(i) What is the name of the gas X left after this process has been
carried out?

(ii) What is the name of the gas Y which dissolved in water?

(iii) Write the equation for the reaction of water gas with steam under
the conditions mentioned. (3)

e The gas X left in d can be mixed with one of the gases in the air and
then used in the preparation of yet another very important industrial
gas; what gas is this? (1)

86. Sulphuric Acid and Carbon Monoxide

a What would you *notice* if some concentrated sulphuric acid were
carefully added to test tubes you were holding which contained

(i) water,

(ii) sugar,

(iii) a crystal of copper sulphate? (3)

b Carbon monoxide is produced when concentrated sulphuric acid is
added to methanoic (formic) acid, HCOOH. Write the equation for this
reaction. (1)

c (i) How is carbon monoxide best produced in the laboratory from
carbon dioxide?

(ii) Draw a sketch of the apparatus you would use (no details needed
for the preparation of the dioxide).

(iii) Write the equation for this reaction in which the monoxide is
produced. (3)

d Another organic acid from which carbon monoxide can be produced
is ethanedioic acid, more often known as oxalic acid, $(COOH)_2$. Explain
how you would prepare a reasonably *pure* sample of the dry gas from this
acid, giving brief details only and no equation. (1)

e Explain why carbon monoxide is a dangerous gas and how it acts in this
connection. (2)

87. The Nitrogen Cycle

In industry, nitrogen from the atmosphere is fixed by means of the Haber
process, which requires extreme conditions of temperature and pressure.
In nature, on the other hand, bacteria are able to fix nitrogen in the soil

at a temperature of about $10° - 30°C$ and normal atmospheric pressure in an aqueous environment. A modern version of the nitrogen cycle is shown below.

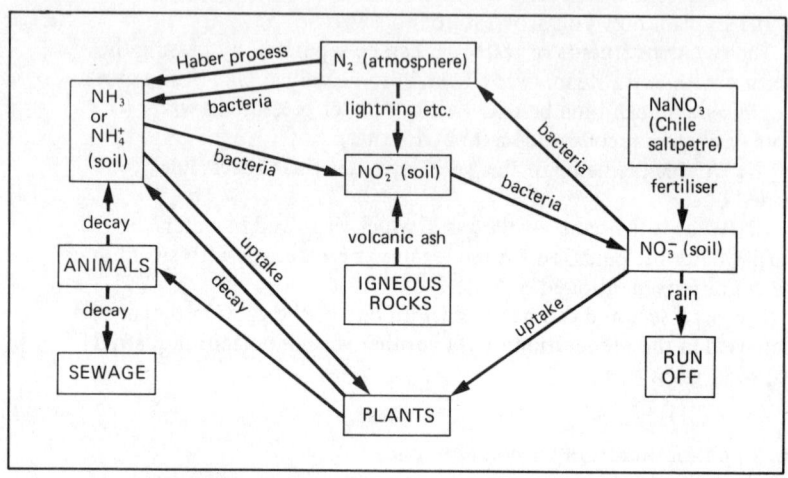

(Adapted from the *New Scientist*, 19 Feb. 1976.)

a (i) Give the name of one *type* of compound found in animals which is rich in nitrogen.

(ii) Give the name of one compound which is used as an *artificial* fertilizer (i.e. a man-made product). (2)

b Write a balanced equation for the reaction which takes place in the Haber process. (1)

c In the Haber process, a pressure of $200 - 1000$ atmospheres is commonly used. Explain, giving your reasons, the effect that such a high pressure has on

(i) the rate of the reaction,

(ii) the position of equilibrium. (2)

d The formation of ammonia is exothermic, but in the Haber process a temperature of about $500°C$ is commonly employed. Explain, giving your reasons, the effect this temperature has on

(i) the rate of the reaction,

(ii) the position of equilibrium. (2)

e Recent chemical research has attempted to find catalysts which will allow nitrogen to be converted to ammonia under approximately the same conditions as bacteria use. Suggest why such a route would be an advantage over the Haber process. (1)

f What effects has man had on the Nitrogen cycle, and what long-term consequences may this have? (2)

88. Fertilisers and Economics

Three common nitrogenous fertilisers are sodium nitrate ($NaNO_3$), cyanamide ($CaCN_2$) and urea ($CO(NH_2)_2$).

a Calculate the mass of one mole of each fertiliser and then the percentage of nitrogen in each;
  (i) sodium nitrate
  (ii) cyanamide
  (iii) urea   (3)

    Two well-known plant fertilisers contain different amounts of available nitrogen:
    brand **X** contains 30% of cyanamide and costs 70 p per kg.
    brand **Y** contains 20% of urea and costs 60 p per kg.
b Calculate the actual mass in grams of the nitrogen present in 10 kg of
  (i) fertiliser (**X**)
  (ii) fertiliser (**Y**).

    A market gardener wishes to apply a fertiliser to a certain patch of ground to provide a total of 2 kg of nitrogen.   (2)
c What mass in kg (to one decimal place) of
  (i) fertiliser **X**
  (ii) fertiliser **Y**
  would provide this amount of nitrogen?   (2)
d Which of the two fertilisers, brand **X** or brand **Y**, would be the more economical to use? (Show your working.)   (1)
e Good all-round fertilisers contain two other important elements (combined, of course) in addition to nitrogen. Give EITHER the chemical names of both elements OR the common names by which gardeners and horticulturists refer to them.   (2)

89. Oxidation and Reduction of Sulphur Compounds

The reaction represented by the equation below is one which you have probably not met before. Study it, and then try to answer the questions which follow.

$$4Na_2SO_3(s) \longrightarrow 3\,Na_2SO_4(s) + Na_2S(s)$$
    A   **B**   **C**

a Give the name of each substance occurring in the equation.   (2)
b Each anion present has the same charge; what is this?   (1)
c Which one of the two substances, **B** or **C**, is formed by the reduction of **A**?   (1)
d Which one of the three compounds would not be affected by the addition of 2 M hydrochloric acid?   (1)
e (i) Which of the three compounds, when treated with 2 M sulphuric acid, gives off a gas which decolorises aqueous potassium manganate(VII)

(potassium permanganate) and leaves a clear colourless solution.

(ii) In this reaction, does the gas evolved behave as an oxidant or as a reductant with the potassium salt? (2)

f (i) If you wished to prepare crystals of **A** in the laboratory, which two compounds, other than water, would you use as starting materials?

(ii) Write a balanced equation for this reaction. (2)

g Aqueous barium chloride will form a precipitate with solutions of two of the substances, **A**, **B** and **C**, but one of these dissolves on the addition of 2 M hydrochloric acid. Give the name and formula of the precipitate which does NOT dissolve in this acid. (1)

Section 13
Carbon Chemistry

90. Petroleum

Crude petroleum (crude oil) is a mixture of many hydrocarbons.

a In what manner, according to scientists' belief, was petroleum originally formed? (1)

On arrival at an oil refinery, the petroleum is split into a mixture of simpler molecules.

b What method is used to achieve this separation into simpler mixtures? (1)

Three typical mixtures resulting from this separation could be:
A hydrocarbons from C_4H_{10} to $C_{12}H_{26}$ (gasoline: used for making petrol).
B hydrocarbons from C_9H_{20} to $C_{16}H_{34}$ (kerosine: used for making domestic paraffin).
C hydrocarbons from $C_{15}H_{32}$ to $C_{25}H_{52}$ (gas oil and diesel oil).

c (i) In which of the mixtures, **A**, **B** or **C**, would the individual components have the highest boiling points?

(ii) More than one structural formula can be written for each of the molecules whose formulae are given above. What word is used to describe compounds with the same molecular formula but different structural formulae? (2)

Mixtures such as **C**, above, may undergo a further process in the oil refinery; this is known as catalytic cracking. In this process, the hydro-carbon mixture is mixed with a very finely divided catalyst (silicon dioxide + aluminium oxide) at about 500°C, when the cracking reaction takes place. Typical products include carbon, unsaturated hydrocarbons such as C_2H_4, saturated hydrocarbons and hydrogen.

d (i) Most of the hydrocarbons produced in this cracking process are liquids at room temperature and pressure. How can this complicated mixture be separated?

(ii) Suggest a reason why it may be advantageous to crack a mixture such as **C** rather than a mixture such as **A**.

(iii) State one large scale (industrial) use for the hydrogen produced in this process.

(iv) State one large scale (industrial) use for any one of the unsaturated hydrocarbons produced in this process.

(v) The carbon produced in the cracking tends to form a coating on the catalyst particles. It has to be removed or the activity of the catalyst is reduced. Why would the catalyst's activity be reduced by the carbon coating?

(vi) The carbon coating is removed by burning it off in air, which heats the powder to about $600°C$. Far from being a nuisance, this is very useful. Why is it useful to have the catalyst heated in this way?　　(6)

91. Alkanes and Alkenes

Hydrocarbons contain only two elements, carbon and hydrogen. One class of hydrocarbons is called the alkanes and the general formula for these compounds is C_nH_{2n+2} .

a (i) Give the name and formula of the third member of the alkane series.

(ii) Give the formula of the member of this series which has 29 atoms in its molecule.　　(2)

b Another series of hydrocarbons is called the alkenes. What is the general formula for this series?　　(1)

c (i) What is the general name given to series like these in which the difference in the total number of carbon and hydrogen atoms between any two successive members is always the same?

(ii) Write the 'formula' for this regular difference.

(iii) What would be the difference in relative molecular mass between the first and fourth members of (1) the alkane series, (2) the alkene series?　　(3)

d (i) The butane molecule can exist in two forms, both of which have the same molecular formula, but different arrangements of the atoms; what are these different forms called?

(ii) Draw, or write out, the structures of the two different forms of butane.　　(2)

e (i) If you had to prepare the first member of the alkane (or alkene) series of hydrocarbons in the laboratory from a named organic compound, which compound would you choose? State which gas you are preparing.

(ii) Write the equation for its preparation.　　(2)

92. Fuels of Different Kinds

a The ignition temperature of a substance is the lowest temperature it must reach before it will burn. A similar term used in connection with liquids only (although not quite the same) is 'flash point'. Arrange the following liquids in the order you would expect for their flash points, putting the lowest one first:

    (i) diesel oil,

    (ii) petrol,

    (iii) kerosine (paraffin). (2)

b Why does a lighted match thrown on to a jar of kerosine often go out, yet a piece of paper soaked in the same liquid inflames very easily? (1)

c A solid fuel fire usually burns better at the base of a tall chimney than with a short one because a better draught is produced and more air is drawn in. Explain briefly, but clearly, exactly why more air is drawn through the fire with a tall chimney than through one which has a short chimney, i.e. why more draught is produced. (2)

d What is the name of the compound we call natural gas or North Sea gas?

    (ii) Write a balanced equation for the burning of natural gas. (2)

e When magnesium burns it produces far more heat than the same mass of wood or coal does, but it is not such a good fuel: why do you think this is the case (ignore cost and the availability of the substances)? (1)

f If the molar heat of combustion of ethanol (C_2H_5OH) is 1370 kJ, (this is a 'practical' question) how much heat would be given out by completely burning 1·15 kg of the fuel? (2)

93. Fermentation

Some glucose ($C_6H_{12}O_6$) solution was placed in a flask with some yeast. The flask was placed in a water-bath at about $30°C$ and a delivery tube from the flask led into some aqueous calcium hydroxide in a beaker. After a short time a gas began bubbling through this solution which was becoming whitish in appearance. After three days the flask was removed from the bath and it was noticed that the calcium hydroxide solution was as clear as at the start.

a To which class of organic compounds does glucose belong? (1)

b What is the purpose of the yeast in this experiment? (1)

c (i) Give the chemical name and formula of the white substance formed in the beaker.

    (ii) Why had the whitish colour disappeared three days later? (2)

The contents of the reaction flask were filtered and 100 cm³ of the filtrate transferred to another flask set up for distillation with a thermometer to indicate the temperature and a receiver below the end of the condenser. As the temperature was raised, a colourless liquid started dripping from the condenser when the temperature in the flask was about $90°C$. At $94°C$ the receiver was changed and a separate portion collected.

This was done again at 96°C and also at about 100°C. The temperature then remained steady until almost all the liquid in the flask had distilled over. The volumes collected in the three receivers were analysed with these results:

Sample	Temperature °C	Volume/cm³	Volume of ethanol in sample/cm³
A	90	16	8
B	94	8	2
C	96	16	2
D	100	60	trace only

d (i) What was the percentage of ethanol in the total volume collected?
   (ii) What was the percentage of ethanol in the sample most likely to burn?
   (iii) Write an equation for the combustion of ethanol in air.
   (iv) Give the name of the liquid in sample **D**.   (4)
e (i) Draw a simple sketch of a device which could be fitted to the neck of the distillation flask (between it and the condenser) to make the distillation more efficient.
   (ii) What is this piece of apparatus called?   (2)

94. Ethanol

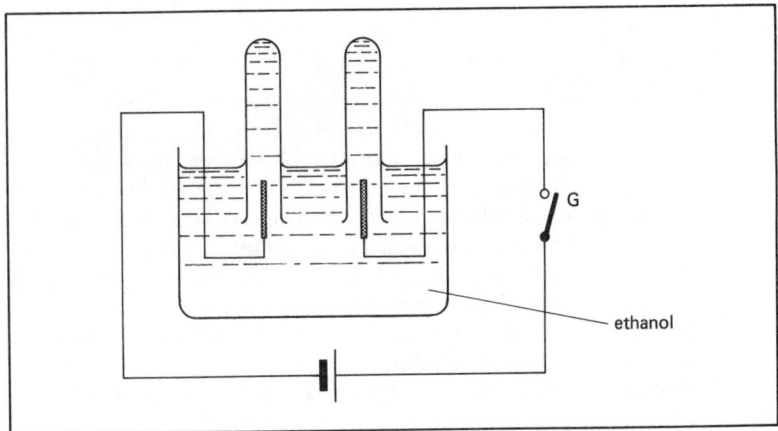

a Ethanol (C_2H_5OH) is the second member of an alcohol series of organic compounds. Give the name and formula of the first member of this series.   (1)

b What would you notice if the circuit in the diagram above was completed by closing the switch G? (1)

c Some ethanol is placed in a test-tube and mineral wool added to soak up the liquid. This is then strongly heated in a horizontal position and a gas is obtained which burns with a very yellow flame. Give the name and formula of this gas. (1)

d When ethanol is heated with concentrated sulphuric acid and solid potassium dichromate(VI), the orange colour of the dichromate(VI) soon disappears and a green solution is left.

(i) What does the production of this green solution from the dichromate(VI) tell us about the kind of reaction that is occurring?

(ii) Give the name OR formula of the organic substance formed from the ethanol. (2)

e If the ethanol is heated with concentrated sulphuric acid and potassium manganate(VII) (potassium permanganate) for some time it is converted into an organic acid.

(i) Give the name and formula of this acid.

(ii) What is the main type of bonding present in this acid? (2)

f If this organic acid is mixed with ethanol and a little concentrated sulphuric acid and heated, a sweet-smelling substance is produced. To what class of organic compounds does this new substance belong? (1)

g When ethanol reacts with phosphorus trichloride a substitution product is formed.

(i) Which element or group of elements in ethanol is replaced by chlorine?

(ii) What is the relative molecular mass of the new compound formed? (2)

95. Soap

Soaps can be made by boiling a fat or an oil with sodium hydroxide pellets and some ethanol. The ethanol acts as a solvent for the other two substances, but does not otherwise enter into the reaction. If the fat contains glyceryl stearate (an ester) the products are sodium stearate (the soap) and glycerol, in which it is soluble. The products are separated by 'salting-out', i.e. by pouring the mixture into brine.

a (i) Give the formulae for the two constituents of brine.

(ii) Ethanol and glycerol both belong to the same class of organic compounds; give the name of this class.

(iii) Which particular property of soap makes it possible for it to be obtained by 'salting-out'? (3)

b If, for brevity, we represent glyceryl stearate by $GlSt_3$ and glycerol by $Gl(OH)_3$, (these are not actual formulae), write the equation for the reaction of glyceryl stearate with sodium hydroxide. (1)

c (i) Give the name of any compound which, if present in solution in water, makes the water hard.

(ii) State the type of hardness produced by this compound. (2)

d (i) What is the chemical name of the whitish precipitate (scum) formed when soap is added to hard water?

(ii) Write the equation for the reaction in solution of the compound you have named in (c) with soap, using the formula for soap derived in (b). (2)

e Soap is soluble in distilled water. How do you think the molecules of soap compare in size with those of water? Are they larger, smaller or about the same? (1)

f In what way could you EASILY show the difference between a solution of soap and a liquid synthetic detergent? State the result of the experiment you would carry out. (1)

96. Carbohydrates

Carbohydrates (also called saccharides or, simply, sugars) may be simple ones such as glucose, double ones such as sucrose (cane sugar) or complex ones such as starch and cellulose.

a Starch and cellulose contain thousands of similar groups in their molecules: what name is given to such complex organic compounds? (1)

b In these carbohydrates the number of hydrogen atoms in the molecule is always twice the number of oxygen atoms. Give the name of a very much simpler compound which also shows this ratio. (1)

c Starch has a molecular formula $(C_6H_{10}C_5)_n$ where n is an exceedingly large number. One molecule of starch can be converted into n molecules of glucose by boiling the starch with a dilute acid. This has the effect of breaking down the large starch molecules by adding to each molecule of it n molecules of water.

(i) What is the name given to this process?

(ii) Write an equation for this change (using one molecule of starch and n molecules of water). (2)

d (i) Starch can also be changed into glucose by the action of saliva. What *kind* of substance present in saliva brings about this change?

(ii) Give the name of another 'breaking-down' process in which substances similar to those present in saliva play an important part. (2)

e When sucrose is broken down by heating in a tube, one of the elements present soon becomes very noticeable.

(i) Which element is this?

(ii) How do you recognise its presence in this case? (2)

f If concentrated sulphuric acid is carefully added to a concentrated solution of sucrose in a tube, what TWO effects will quickly be noticed?(2)

97. An Investigation of the Breakdown of Starch

Starch and cellulose are both carbohydrates, but whilst human beings can use starch as a food, they cannot use cellulose. It was decided to try to investigate the reason for this and eight test tubes were set up as in the table here:

A	starch + water
B	starch + water + 2 M hydrochloric acid
C	starch + water + saliva
D	cellulose + water
E	cellulose + water + 2 M hydrochloric acid
F	cellulose + water + saliva
G	maltose + water
H	glucose + water

The test tubes were kept at a temperature of 37°C. From time to time, test tube C was tested to see if there was any starch left. When none remained, a drop of the solution was taken from each of the test tubes and placed along a line drawn near the bottom of a piece of chromatography paper (good quality filter paper). The paper was then suspended with the bottom dipping into a mixture of propanol, acid and water for several hours. It was removed, allowed to dry, and then dipped briefly into a solution of several chemicals. When the paper was warmed gently, coloured spots became visible, as show in the diagram.

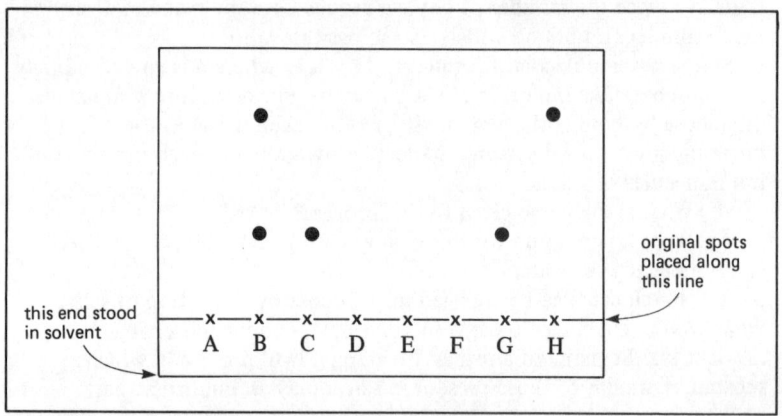

a   Explain the meaning of the word 'carbohydrate'. (1)
b   How would you test the contents of tube C to find out if any starch remained? (1)
c   Suggest a possible reason why 37°C was chosen as the temperature at which the test tubes were to be kept. (1)
d   (i) What was the purpose in standing the chromatography paper in the mixture of propanol, acid and water?
    (ii) Most of the chemicals used in this experiment are colourless: How, then, were they rendered visible for the purpose of identification?(2)

e What purpose was served by tubes A, D, G and H? (1)

f What product(s) is/are formed when starch is broken down by
  (i) saliva
  (ii) 2 M hydrochloric acid? (2)

g (i) What conclusion can be drawn from the absence of a spot with tubes **E** and **F**?

  (ii) Suggest a reason why cellulose is no good as a food for human beings, whereas starch is. (2)

98. Home Winemaking

The following is a recipe for making wine at home: 'Take about 3 lb of sugar and dissolve it in warm water. Add some rose petals to give it flavour, and enough water to make the final volume up to 1 gallon. Add yeast and allow the mixture to ferment in a warm place for some weeks. The mixture must be kept in a container which is not sealed, but at the same time it must be impossible for air to get in.'
For the purpose of this question take 3 lb = 1368 grams and 1 gallon = 5 dm^3 (5 litres).

a The sugar referred to in this experiment is sucrose or cane sugar, $C_{12}H_{22}O_{11}$. First of all, the yeast acts with the water to change the sucrose to glucose, $C_6H_{12}O_6$. Write a balanced equation for this reaction (the yeast is acting in a manner similar to that of a catalyst). (1)

b (i) In order to get a good yield of ethanol in the wine, the mixture has to be left for several weeks in a warm place. Suggest why boiling the mixture will probably NOT succeed in speeding up this process.

  (ii) In baking bread, yeast is also used. State why, in that case, unlike in brewing, it is advantageous to use a high temperature. (2)

c After fermentation, the mixture will still contain yeast along with the ethanol formed.
  (i) How could the yeast be removed?
  (ii) How could reasonably pure ethanol then be obtained? (2)

d How many moles of sucrose are used in the recipe given above? (1)

e Apart from ethanol, carbon dioxide is formed when glucose is fermented.
  (i) Write a balanced equation for this reaction.
  (ii) How many moles of carbon dioxide could be obtained by the fermentation of all the glucose obtained from the 1368 grams of sucrose? Show your working.
  (iii) Calculate the volume of carbon dioxide gas (measured at room temperature and pressure) which would be liberated during the fermentation of the 1368 grams of sucrose, and hence point out why the instructions state that the container must not be sealed. (4)

99. Artificial Polymers and Fibres

a Give the name of
  (i) a polyamide
  (ii) a polyester   (2)
b Give the names of ALL the elements present in polystyrene.   (1)
c Sketch the repeating unit in polythene.   (1)
d A certain polymer is composed of many repeating units of the form

$$\left[-\; \begin{array}{l} \qquad\quad CH_3 \\ \qquad\quad | \\ CH_2-C \;-\!-\!- \\ \qquad\quad | \\ CH_3-C=O \end{array} \;- \right]$$

  (i) What type of structure does this polymer have, i.e. what functional group is present and repeated?
  (ii) What would be the physical state of this polymer at room temperature?   (2)
  The block diagram below represents part of the structure of nylon, the only elements present being carbon, hydrogen, nitrogen and oxygen. Details of the blocks are not shown, nor are they required for this question. All bonds are represented by single lines although some of them are *double* bonds.

$$-\left[\begin{array}{c} -\,A\,-\,\square\,-\,E\,-\,J\,-\,\square\,-\,M\,-\,R\,-\,\square\,-\,X\,- \\ \quad\; |\qquad\qquad |\quad\; |\qquad\qquad |\quad\; |\qquad\qquad |\quad \\ \quad\; D\qquad\qquad G\quad L\qquad\qquad Q\quad T\qquad\qquad Z\quad \end{array} \right]-$$

e (i) If there are only TWO atoms of carbon represented by letters, which are these two letters?
  (ii) List all the letters representing atoms of nitrogen.
  (iii) How many hydrogen atoms are represented by letters in the diagram?
  (iv) Which bonds are double bonds? Give the letters at the ends of the bonds.   (4)

100. Decomposition of a Plastic

A student was investigating a sample of a plastic. It was found that when the plastic was heated in the absence of air it decomposed to give a gaseous product, insoluble in water, and also a product which condensed easily to form a liquid.

On shaking the gas with bromine water, the brown colour of the bromine was rapidly removed and the volume of gas decreased. On using a large excess of bromine water, about three-quarters of the gas was consumed, but the remainder was unaffected, no matter how much bromine was used. When this remaining gas was ignited, it burned with a bluish flame, but NO carbon dioxide was produced. The product which condensed as a liquid did NOT decolorise bromine water. It did, however, burn easily with a smoky flame, producing carbon dioxide and water.

a Draw a diagram of the apparatus you would use to heat the plastic in the absence of air, and to collect samples of both the gaseous and condensable products. (3)

b (i) What can be deduced from the fact that the gas decolorised bromine water?

(ii) Suggest a reason why the volume of gas decreased on shaking with bromine water?

(iii) Why did some gas remain after shaking with even a large amount of bromine water?

(iv) What can be deduced from the fact that the gas which remained after treatment with bromine water did NOT produce carbon dioxide on combustion?

(v) Suggest what this remaining gas is likely to be. (5)

c What can be deduced about the chemical nature of the condensable product? Explain your reasoning. You are not required to attempt any identification. (2)

101. Polymers

Compounds **A** and **B** may be reacted together to form a polymer.

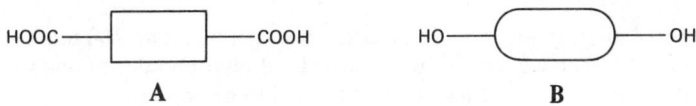

| A | B |

a (i) What type of compound is **A** (what functional group)?

(ii) What type of compound is **B** (what functional group)?

(iii) What type of compound (what functional group) is formed when one molecule of compound **A** reacts with one molecule of compound **B**? (3)

b (i) Use block diagrams to indicate the structure of the polymer formed by compound **A** and **B** reacting together.

(ii) Explain why a polymer would NOT form from a reaction between compound **A** and compound **C**, where **C** has the following structure: (2)

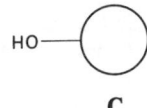

C

c Give one example of a polymer formed by the *type* of reaction which occurs between compunds **A** and **B**, although not necessarily involving the same functional groups.   (1)

d Compound **D** will polymerise by itself under suitable conditions.

$$
\begin{array}{ccc}
H & & H \\
| & & | \\
C & = & C \\
| & & | \\
H & & X
\end{array} \quad D
$$

(i) Draw a diagram to show the structure of the polymer formed from compound **D**, (give three units).

(ii) By what *type* of reaction does compound **D** polymerise?

(iii) The symbol, X, in compound **D**, could represent any one of a number of atoms or groups of atoms. Draw the structural formula of one possible compound, **D**, and state how it may be made, either in the laboratory or in industry.   (4)

Section 14
Source of Materials and Energy

102. Air and its Gases

A sample of air at room temperature and pressure (20°C and 760 mm pressure) was taken and experiments carried out on it in order to remove the various constituents which analysis showed to be present.

A large evacuated flask with two tap connections was weighed. One tap of the flask was then attached to a supply of air which entered and completely filled the flask at atmospheric pressure. The tap was closed and the flask reweighed. Both taps were then opened and water was run through one and out of the other until the flask was full. Again it was weighed.

mass of evacuated flask	=	273·664 g
mass of flask plus air	=	274·960 g
mass of flask plus water	=	1273·664 g

a (i) What mass of air was originally taken?

(ii) What volume of air at room temperature and atmospheric pressure was taken into the flask?

(iii) Calculate the density of the air in g cm^{-3}

(iv) If hydrogen, under the same conditions as the air, has a density of 0.000 09 g cm^{-3} what is the relative density of air compared with hydrogen? (4)

A fresh sample of air of 1 dm^3 volume at r.t.p. was then passed through a dry U-tube containing silica gel coloured blue with a cobalt salt. The mass of the tube and gel increased and the blue colour changed to pink. Heating the gel restored the blue colour and a vapour came off which condensed to a colourless liquid of mass 10 g.

b (i) What substance was extracted from the air by the silica gel?

(ii) What would be the volume at room temperature of the liquid obtained by heating the gel? (2)

After passing through the silica gel, the gas was next bubbled through aqueous potassium hydroxide, increasing the mass of the solution by 0·3 g.

c (i) What gas was absorbed by the potassium hydroxide solution?

(ii) Write an equation for ONE reaction that occurs here. (2)

The remaining gas, after leaving the hydroxide solution, was passed through a combustion tube in which copper clippings were being heated.

These changed colour to black and the gas emerging from the exit only measured about 800 cm^3 at r.t.p.

d (i) Give the name and formula of the substance into which the copper changed.

(ii) What is the name of the gas likely to be collected at the end of the experiment, and how many atoms does this gas have in its molecule?

(2)

103. An Uncommon Metal

Illmenite is a mineral which contains both iron(III) oxide and titanium(III) oxide; it is an industrial source of the metal titanium. The symbol for titanium is Ti.

a (i) Write the formulae for the two oxides mentioned above.

(ii) If the simplest formula (empirical formula) for ilmenite can be written as $FeTiO_3$, in what molar proportions are the two oxides present? Show how you arrive at your answer. (2)

b Titanium has a more common oxide, titanium(IV) oxide.

(i) Write the formula for this oxide.

(ii) If the titanium was present in a different mineral in the form of titanium(IV) oxide and there was the same number of moles of this oxide as of iron (III) oxide in any sample of the mineral, what would be the simplest formula for this mineral? (2)

c Any sample of a mineral always contains a number of other substances besides the ones mentioned. Analysis of a sample of ilmenite shows that there is 5·6% of iron(III) oxide and 8·4% of titanium(III) oxide present.

(i) What mass of each is actually contained in 500 kg of this ore?

(ii) If the relative atomic masses of iron, titanium and oxygen are 56, 48 and 16 respectively, what are the maximum masses of iron and titanium (separately) which could be extracted from 500 kg of the ore? (4)

d In the industrial treatment of ilmenite to obtain the oxides and to reduce them, in turn, to the metals, the practical processes cause some losses and the final amounts are 10% less than the ones expected for iron and 15% less for titanium. What masses of iron and titanium would *actually* be obtained from 500 kg of ilmenite? (2)

104. The Electrochemical (Activity) Series

Only a few metals are found native on the earth. To some extent it is those metals which have been known longest to man: native copper was used as long ago as 7000 BC in Persia. By about 5000 BC man had learned how to extract copper from malachite, its ore. Although a little iron occurs native, it was not until about 2500 BC that it was extracted from its ores. Aluminium does not occur native at all, and the metal was not isolated until 1825, although it is the third most abundant element (combined) in the earth's crust. In 1862 the price of aluminium was about £600 per kilogram; it is now about £0.3 per kilogram.

a Explain the meaning of the terms
  (i) native
  (ii) ore
  as used in the above passage. (2)
b Suggest why aluminium never occurs native, iron rarely, but copper is relatively common. (1)
c (i) What *type* of reaction is used to extract copper and iron from their ores?
  (ii) Suggest why iron was not extracted from its ores until several thousand years after copper. (2)
d (i) Aluminium is the third most abundant element in the earth's crust; which are the first and second most abundant elements?
  (ii) Why, if aluminium is so common in the earth's crust, was it not extracted until the last century?
  (iii) What method is now used to extract aluminium from its ores?
  (iv) Suggest why the price of aluminium nowadays is so much less than it was 120 years ago. (4)
e On the earth, iron is commonly found as haematite, Fe_2O_3, or other iron(III) compounds. On the moon, on the other hand, the most common iron mineral appears to be iron(II) oxide, FeO. Suggest a possible reason for this difference. (1)

105. Conservation

'Mankind is busy getting oil out of the ground, converting it into carbon dioxide and dumping it in the atmosphere. In the really long-term, atmospheric carbon dioxide may become the best available source for conversion back to hydrocarbons. Admittedly, plant life is already tapping this source of carbon dioxide, but a direct route back to the raw

materials of chemical industry would be preferable. There are two main difficulties in this scheme; first, the atmospheric carbon dioxide is rather dilute, and secondly, it is difficult to convert it into anything more useful. If there were an attractive method of solving the first problem, that is collecting the carbon dioxide, then it would be worth developing a method of pumping energy back into the carbon dioxide in order to convert it into a useful raw material such as methanol'
(Taken from the *New Scientist*.)

a   (i) Which of the following figures is the approximate percentage of carbon dioxide in the atmosphere?
    4%, 0·4%, 0·04%, 0·004%.
   (ii) Calculate the number of litres of carbon dioxide in 10 000 litres of air.   (1)
b   What is the main method by which most of the oil obtained from the ground is converted into carbon dioxide and 'dumped in the atmosphere'?   (1)

c   Oil is a mixture of hydrocarbons and therefore does not have a definite fixed formula. Assuming a formula of C_8H_{18} for ONE of the hydrocarbons in this mixture, write a balanced equation to represent its conversion into carbon dioxide by the method you have given in (b).   (2)
d   (i) What use does plant life make of atmospheric carbon dioxide?
   (ii) What is the substance present in certain parts of green plants which enable them to make use of this carbon dioxide?
   (iii) What other substance is also taken in by plants and plays an essential part in this reaction with the carbon dioxide?
   (iv) When this other substance and the carbon dioxide react together in the body of a plant, what type of organic compound do they form?   (4)
e   The formula for methanol is CH_3OH. Compare this formula with those of the oxides of carbon, and suggest a substance which could be considered for use with ONE of these oxides (stating which one) to produce methanol.   (2)

106. Tomorrow's World

Read the following extract through carefully and then answer the questions which follow.

'Dr. D. Whitten and his colleagues have developed the first artificial chemical compound that can split water into hydrogen and oxygen simply by *trapping the visible light from the sun.* Hydrogen is frequently proposed as a *pollution-free fuel,* but so far there are no commercially viable techniques for generating the gas. . . Producing hydrogen from the relatively limitless supplies of water and sunlight is obviously attractive, especially as a replacement for natural gas, the first of the natural energy supplies likely to run out: . . . The compound is based on the *transition metal* ruthenium . . . This structure is very reminiscent of the active centre of chlorophyll . . . Recently, a ruthenium compound was developed to convert atmospheric nitrogen to ammonia, a process which normally can be achieved only by the plant *enzyme* nitrogenase, or by *industrial processes* demanding *heroic* use of *high temperatures and pressures.* Ammonia is, of course, an important *basic* material for *fertilisers.'* (Taken from the *New Scientist,* July 1976.)

a Why is hydrogen described as a *pollution-free fuel*? (1)

b Ruthenium is described as a *transition metal.* Give the name of a very common transition metal which, when heated, is able to split steam into hydrogen and oxygen, combining with the latter. (1)

c (i) It is said that this new ruthenium compound traps the *visible* light from the sun: how then, does it differ from chlorophyll in its action?

(ii) Give the name of the process in which chlorophyll plays an important part. (2)

d Explain briefly what an *enzyme* is and does. (1)

e (i) What do you think is meant by the term *heroic* in this passage?

(ii) What is the name of the *industrial process* referred to here?

(iii) Give, approximately, the pressure and temperature used in this process. (3)

f (i) In the last line, *basic* means 'fundamental'; what other (chemical) meaning can be attached to this word when used in connection with ammonia?

(ii) Give the name and formula of the ammonium compound which is used in vast amounts as a *fertiliser.* (2)

107. Making a Fuel Cell

The apparatus shown in the following diagram was set up. Both electrodes were made of carbon and the electrolyte was aqueous sodium hydroxide.

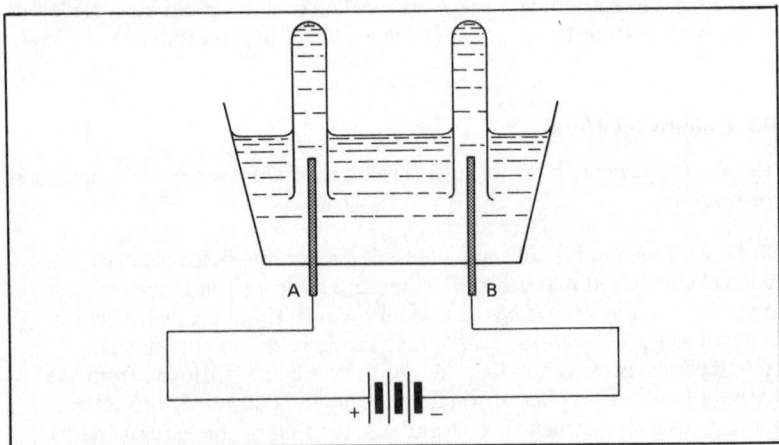

The experiment was carried out in two stages. In the first stage, the two electrodes were connected by means of a battery as shown in the diagram.

Bubbles of gas were seen and collected in the tubes held above A and B. The changes which take place in this part of the experiment may be summarised thus:

$$OH^- \longrightarrow H_2O + O_2 + e^- \qquad (X)$$
$$H_2O + e^- \longrightarrow H_2 + OH^- \qquad (Y)$$

a (i) Balance equations X and Y.
   (ii) At which electrode, A or B, will reaction X take place? Give a reason for your answer.
   (iii) What volume of gas would you expect to be produced at electrode A for every 10 cm^3 of gas produced at electrode B?
   (iv) What can you say about the concentration of the sodium hydroxide in the cell, as the experiment proceeds?   (4)

When sufficient gas had been collected in the two tubes to ensure that each gas was in contact with the carbon electrode, the battery was removed and replaced by a voltmeter. This gave a reading of 1·2 V.
b (i) Write the equations for the reactions now taking place at the two electrodes which give rise to this voltage.
   (ii) Write an overall equation for the net reaction taking place in the cell under these conditions.
   (iii) How many moles of electrons need to be transferred through the circuit to give rise to the formation of 1 mol H$_2$O?
   (iv) What energy is produced by the cell when 1 mol H$_2$O is formed in it? (Assume that 1 volt = 1 joule coulomb^{-1} and that 1 faraday = 1 mol e$^-$ = 96 000 coulombs.)   (5)
c The device made in the second part of the experiment is known as a fuel cell. Suggest one advantage that a fuel cell might have over more conventional ways of generating electricity by burning a fuel and using the heat to raise steam for driving turbines.   (1)

Section 15
Industrial Processes

---

108. Sulphuric Acid

A very important industrial reaction can be represented by the equation:

$$2SO_2(g) + O_2(g) \rightleftharpoons 2SO_3(g) \quad \Delta H = -192 \text{ kJ mol}^{-1}$$

This is the second of three stages in the production of sulphuric acid from sulphur.

a Write the equation for the first stage using elemental sulphur. (1)

b What is the industrial name for the above process by which sulphuric acid is made? (1)

c (i) What does the sign $\rightleftharpoons$ in the above equation mean?

(ii) What does ΔH represent?

(iii) Is the reaction exothermic or endothermic? (2)

d Pure oxygen is expensive to buy and to transport to the site where this operation is taking place. What could be used in its place and why would it be suitable? (1)

e Suggest TWO means by which the rate of production of sulphur trioxide might be increased. (1)

f The sulphur trioxide is converted into sulphuric acid by passing it through concentrated sulphuric acid already produced.

(i) Why is this method used instead of absorbing the trioxide in water as shown by the equation:

$$SO_3(g) + H_2O(l) \longrightarrow H_2SO_4(aq) ?$$

(ii) What is the name of the substance obtained from sulphur trioxide plus sulphuric acid?

(iii) How is this substance finally converted into sulphuric acid of the required concentration?

(iv) Give the balanced equation for the reaction in (iii). (4)

109. Ammonia – The Haber Process

In the preparation of ammonia in industry the raw materials are nitrogen and hydrogen. These are mixed together in the correct proportions needed to form ammonia.

a Write the balanced equation for the manufacture of ammonia, indicating whether the reaction goes to completion or not. (1)

b (i) State briefly how the nitrogen is obtained.

(ii) Give one industrial source from which the hydrogen is obtained.

(iii) How much hydrogen would there be in 400 dm^3 of the gaseous mixture required for the manufacture of ammonia? (3)

c The formation of ammonia from its elements is an exothermic one: if no other factors were involved, do you think the formation would be favoured by a medium or a high temperature? (1)

d (i) If all the nitrogen and hydrogen used were converted into ammonia, how would the volume of ammonia formed compare with the volume of the original mixture?

(ii) Would the production of ammonia be favoured by a high or a low pressure? (2)

e It is found that the rate of reaction can be increased by passing the mixed gases over some finely-divided iron; what effect would this have on the total volume of ammonia obtainable from a fixed volume of the mixed gases? (1)

f The ammonia produced is removed from any unchanged gases in one of two possible ways. Which two physical properties of ammonia enable it to be removed in these ways? (2)

110. The Production of Sodium Hydroxide

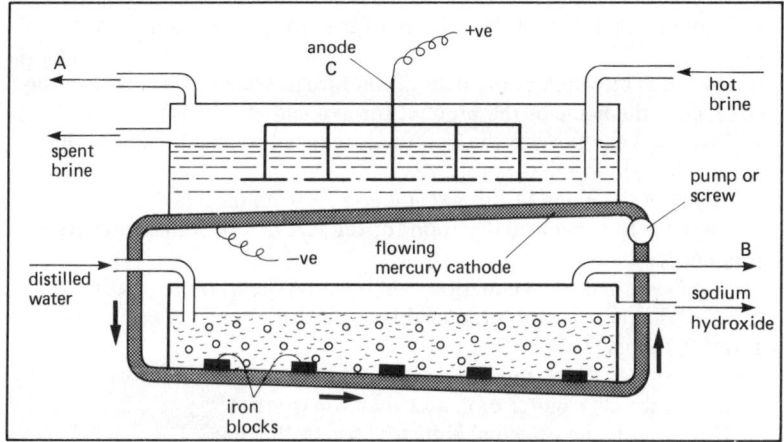

The apparatus shown represents the Kellner-Solvay process for the production of sodium hydroxide industrially. It take place in two stages and involves the electrolysis of brine. In the upper cell the cathode is provided by a stream of mercury flowing along the bottom and this alters the normal course of electrolysis by allowing sodium to be liberated.

a (i) What is brine?
(ii) Give the names and correct formulae (with charges) of ALL the ions present in brine. (3)

The sodium liberated is taken up by the mercury and the substance produced is then pumped to the lower cell where it reacts with the water.
b (i) What product is liberated at A in the upper cell?
(ii) What material is used for the anode, C, in the upper cell? (2)
c (i) What is the name of the substances formed by the sodium and the mercury?
(ii) What product is liberated at B in the lower cell? (2)
d If the sodium-mercury substance is represented by Na/Hg, show by an equation how it is formed from the sodium *ions* set free in the upper cell. (1)

e Write a balanced equation to show how the Na/Hg substance reacts with water in the second cell. (1)

f In another industrial method, sodium hydroxide is produced by the electrolysis of brine in a single cell with an iron cathode. *Explain* how the solution formed is one of sodium hydroxide. (1)

111. The Extraction of Iron

The basic materials needed for the production of iron in the blast furnace are limestone, coke and air in addition to the iron ore.

a Name correctly and fully one ore of iron (or give its name as a mineral.) (1)

b Hot air is blown in at the base of the furnace where it reacts with the coke. Give the name of the product formed and give its formula. (1)

c Higher up the furnace the iron ore is reduced to iron by one of the gases produced in the furnace.

(i) Give the name of this gas and give its formula.

(ii) Give the chemical equation for the reaction by which this gas is produced.

(iii) Give a balanced equation to show how the iron ore is reduced. (3)

d (i) What compound, produced from the limestone, takes part in forming the slag?

(ii) What is the chemical name of the new compound it forms?

(iii) Is the slag lighter or denser than the iron?

(iv) What is the principal industrial use of this slag? (4)

e Iron has a melting point of about 1530°C, but the highest temperature in the blast furnace is not much more than 1100°C. Suggest a reason why the iron produced in the furnace becomes sufficiently molten to be run off. (1)

Section 16
Miscellaneous Items

112. A Simple Reaction

When 2 M sulphuric acid is added to a solid **A**, a gas **B** is produced which causes a white precipitate to form on passing it through aqueous calcium hydroxide. If some of the gas **B** is passed into water containing universal indicator, the indicator changes colour from green to a yellow-orange, but this change is reversed when the solution is heated. If some magnesium ribbon is lit and placed in a jar of the gas **B**, it continues to burn with a crackling noise, forming copious white fumes **C**. These fumes dissolve when dilute sulphuric acid is added and small black particles are noticed floating in the solution **D**. The solid **A** is soluble in water and does not decompose when heated in a Bunsen flame.

a   (i) Give the name and formula of the gas **B**.
    (ii) Which of the above reactions indicates that the gas **B** probably contains oxygen?
    (iii) What are the small black particles present in solution **D**?   (3)
b   (i) Is **B** an acidic or alkaline gas?
    (ii) Which reaction shows this?
    (iii) When the solution of the gas is heated, does its pH increase or
decrease?   (3)
c   (i) Write a balanced equation for burning magnesium in gas **B**.
    (ii) What would you NOTICE if some aqueous barium chloride
were added to solution **D**?   (2)
d Suggest two possible substances, either of which might be **A**.   (1)
e Is **A** an ionic or a covalent solid?   (1)

113. Heat on a Variety of Compounds

The following is a list of compounds which are solids at room temperature.

calcium carbonate	copper(II) nitrate	mercury(II) oxide
calcium oxide	hydrated copper(II)	sodium hydrogen-
copper(II) carbonate	sulphate	carbonate
	lead(II) nitrate	sodium nitrate

Using only compounds named in this list, give the name of ONE compound in each case which, when heated, behaves as described below. Any compound may be used more than once if you wish.

a Gives off a brown gas.   (1)
b Leaves a black solid residue.   (1)
c Leaves an orange/yellow solid residue.   (1)
d Forms a metal.   (1)
e Leaves a solid residue which gives off a gas when treated with a
dilute acid.   (1)
f Decomposes, but leaves a solid residue of the same colour.   (1)
g Only decomposes slowly even at a very high temperature.   (1)
h Does not decompose whatever the temperature.   (1)
i Undergoes a change which is reversible.   (1)
j Gives off a gas which can be condensed to a liquid in a freezing mixture.
   (1)

114. The Cooling Curve for Cyclohexane

Some cyclohexane, C_6H_{12}, was poured into a test tube, which was then stood in a beaker containing crushed ice mixed with salt. A thermometer was inserted in the test tube and the temperature noted every 30 seconds. The results are shown on the following cooling curve.

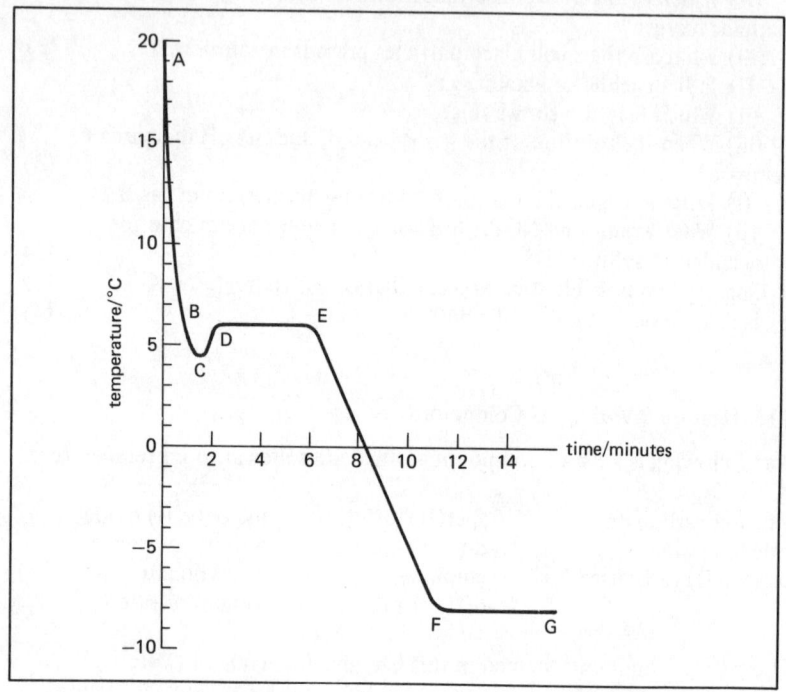

a The little dip in the graph, marked by the letters, B, C, D, is due to supercooling. How would you try to avoid this if you were doing the experiment?   (1)

b Use the cooling curve to answer the following questions:
   (i) What was room temperature on the day of the experiment?
   (ii) What is the freezing point of cyclohexane?
   (iii) What was the temperature of the ice/salt mixture?   (3)

c The cooling curve shows the following features (not necessarily in this order):
   (i) liquid cyclohexane cooling down,
   (ii) solid cyclohexane cooling down,
   (iii) a mixture of liquid and solid cyclohexane,
   (iv) solid cyclohexane at constant temperature.
Use the letters, AB, DE, EF and FG to identify each of these regions.

(3)

d Describe carefully what is happening to the molecules of cyclohexane in the region DE. (2)

e In what way would the cooling curve have differed if crushed ice alone had been used, instead of the ice/salt mixture? (1)

115. Pollution

'Every time it rains in Athens, it rains dilute sulphuric acid and a little bit more of the Acropolis disappears. The ancient marble has suffered more during the last thirty years than it had during the two thousand that preceded it. Ecologists will be familiar with the simple chemical formula. Industrial chimneys, heating and transport fuel, all give off sulphurous fumes which hit the atmosphere as sulphur dioxide ($SO2$).

Mixed with another atom of oxygen and then dissolved by rain (HO), it comes down as dilute sulphuric acid ($HSO4$) and does what the Franks, Venetians and Turks, who all left their mark on the temple of Athens, could never quite manage. It makes the marble crumble.

This type of pollution is a phenomenon to some extent common to all industrialised countries, but since Athens has some of the worst air pollution in Western Europe, it suffers more than most. According to one recent survey, 150 000 tons of sulphur dioxide are pumped into the Athenian air every year.

The [statues] are in a very bad way. If they continue to be exposed to the polluted elements, their features will soon be obliterated. The [statues] are now too far gone to preserve them *in situ*.

An international panel of experts has decided that they can only be saved by removing them to the Acropolis Museum and replacing them with replicas made out of cement and marble dust. The cast for the replicas has already been prepared by the British Museum from the [statue] Lord Elgin removed to England over 100 years ago, which is in much better condition. But taking the figures down without damaging them is a complicated task, and Mr. George Dontas, the Director of the Acropolis, thinks it unlikely that it will be done much before the end of the year. In the meantime, the [statues] will be sealed where they are in glass and steel boxes filled with nitrogen gas, to keep them at a constant temperature.' (Taken from *The Observer*, 15th January 1978.)

a Rewrite the sentence 'Mixed with . . . never quite manage' (lines 7–10) correcting all the chemical mistakes. (2)

b Write properly balanced chemical equations to describe the conversion of sulphur dioxide to sulphuric acid. (3)

c What approximate mass of sulphuric acid could be formed from the 150 000 tons of sulphur dioxide pumped into the Athenian air every year? (1)

d The marble statues crumble because marble is a form of calcium carbonate. Write an equation for the reaction which takes place between this and the sulphuric acid. (1)

e Explain carefully why the marble has suffered more damage in the last 30 years than in the previous 2000. (1)

f Suggest why the statue kept at the British Museum in England for the last 100 years or so is in much better condition than those remaining in Athens. (1)

g Suggest why the boxes used as a temporary measure to protect the statues are to be filled with nitrogen. (1)

116. An Unusual Reaction

Titanium dioxide is a finely-powdered white solid. In a research project, carbon tetrafluoride was heated with excess titanium dioxide in a sealed glass tube. It was expected that carbon dioxide and titanium tetrafluoride would be formed. Some information about the reactants and expected products is given in the following table.

Name	Formula	Melting point/$^\circ$C	Boiling point/$^\circ$C
carbon tetrafluoride	CF_4	-184	-128
titanium dioxide	TiO_2	1840	$2500 - 3000$
carbon dioxide	CO_2	sublimes at -78.5°C	
titanium tetrafluoride	TiF_4	sublimes at 284°C	

After being heated for several days, the tube was cooled to room temperature, opened, and the gas inside transferred to a previously weighed graduated syringe, which was then weighed again. The results were:

volume of gas in syringe at room temperature and pressure = 96 cm^3
mass of empty syringe = 159·732 g
mass of syringe + gaseous products = 160·092 g

a (i) What type of structure is carbon tetrafluoride likely to have? Give a reason for your answer.

(ii) What type of structure is titanium dioxide likely to have? Give a reason for your answer.

(iii) Suggest a reason why the boiling point of titanium dioxide is quoted as $2500 - 3000^\circ$C, rather than as a single sharp figure. (3)

b Write an equation for the reaction which is expected to take place in the tube, including symbols of state for the reactants and products as they would be at room temperature and pressure. (2)

c (i) Use the information given to calculate the mass of 1 mole of the gas collected in the syringe. Show your working in full.

(ii) Could the gas collected in the syringe be either carbon dioxide or unreacted carbon tetrafluoride? Explain how you arrive at your answer. (4)

d It was suggested that the gas collected at the end might in fact contain silicon tetrafluoride. If correct, where would the silicon have come from? (1)

117. The Action of Heat on an Oily Liquid

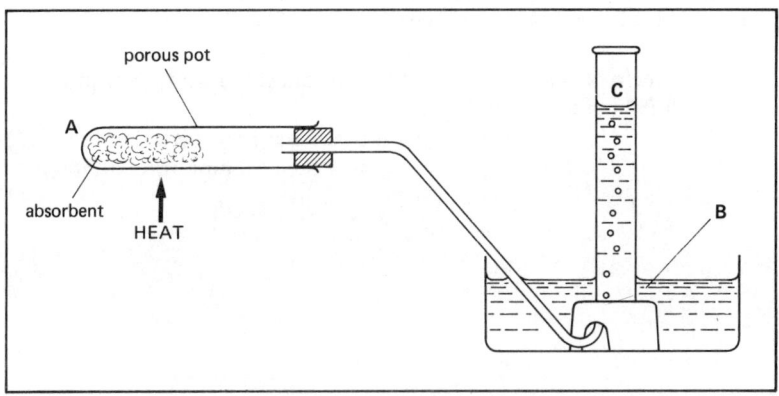

A viscous oily liquid **A**, which was not a solution and had a density nearly twice that of water, was placed in a test tube and wisps of an inert absorbent material added to soak up the liquid. A layer of porous pot was placed on top and the tube set up as shown in the diagram. The porous pot was heated and the liquid vaporised. White fumes came off and these were passed through water to form a solution **B**, while a gas **C** was collected in the jar. No residue remained on the absorbent after all the liquid had disappeared. At the same time, a colourless gas **D** was separately prepared by adding some copper to concentrated nitric acid mixed with an equal volume of water. **C** and **D** were mixed in a jar and formed a brown gas **E**.

a (i) What is the name of the colourless gas **D**?
   (ii) What gas is the brown gas **E** likely to be?
   (iii) What gas must **C** be if it converts **D** into **E**?
   (iv) What would be the effect of a solution of **E** on neutral litmus paper?   (2)
b Write a balanced equation for the reaction between **C** and **D**.   (1)
   Another gas jar of **C** was obtained and into it was poured an alkaline solution of pyrogallol; the jar was covered and shaken vigorously. It was then stood with the mouth under water in a dish and the lid removed.
c Describe and explain what you would see when this was done.   (2)
   The solution **B** in the trough was found to have a pH of about 2 after the experiment; aqueous sodium hydroxide was added to it while stirring until the pH was 7.
d (i) What does a pH of 7 signify?
   (ii) If some neutral litmus solution was added to the original.
solution **B**, what colour would it become?

(iii) What type of compound might have a pH of 2?

(iv) Name ONE compound which, in solution, has a pH of 7.　　(2)

Some of the treated solution **B**, now of pH 7, was added to aqueous barium chloride acidified with hydrochloric acid and a white precipitate was formed.　　(2)

e　(i) Name one anion which, in solution, can give a white precipitate with acidified barium chloride solution.

(ii) Give the correct formula (with the charge) for this anion.　　(2)

f　Now write a balanced equation to show the decomposition of liquid **A** by heat.　　(1)

Answers to numerical items

Section 1 Matter

3 d (i) $0.000\ 1\ cm^3$ (ii) $0.000\ 025\ cm^3$, (iii) $125\ cm^3$
  (iv) $(25 \times 10^{-6})/125 = 2 \times 10^{-7}\ cm$

4 d (i) about 15.2 s (ii) about 29

5 a 0.125 mol CO_2 b (i) $3000\ cm^3$ (ii) 300 atm (iii) 373 atm
  d (i) 0.125 M

Section 2 Atomic Structure

6 c 69.8 8 e 2 f 2

Section 3 Mole Concept

10 a (i) 0.5 g (ii) 2.4 g (iii) C_2H_5 b (i) 57.6 (ii) C_4H_{10}

11 a $1 : 2$ 12 b (i) 28 g c 2 13 d (ii) $327°C$ e $24\ dm^3$

14 f (i) 3.2 g Cu, 0.4 g 0 (ii) 0.05 mol Cu, 0.025 mol 0 (iii) Cu_2O

15 d (i) 0.62 g Pb, 0.003 mol Pb (ii) 0.48 g Br, 0.006 mol Br (iii) $PbBr_2$

16 f (i) 0.71 g CI = 0.02 mol C1, 2.54 g I = 0.02 mol I, ICl

17 c $20\ cm^3$ N_2, $40\ cm^3$ H_2, N_2H_5

18 a (i) 0.0025 (ii) 0.005 b (i) $20\ cm^3$ (ii) 0.25 M c (i) $40\ cm^3$
  (ii) 0.1 d (i) $0.002\ 5$ (ii) 4

19 c (i) $2.5\ cm^3$ (ii) 0.005 mol $(NH_4)_2S$ (iii) 0.005 mol $CdSO_4$

Section 4 Structure and Bonding

26 a $18°C$ b $58°C$ d 0.5 mol $C_2F_3Cl_3$ e (i) 2 minutes (ii) 120 J
  f (i) 5 minutes (ii) $24\ kJ\ mol^{-1}$

Section 5 Electrochemistry

29 b 3860 c 0.02 d 193 000 e 2

30 b (iii) 0.5 mol Cu c (i) 300 coulombs (ii) $1/320$ mol Cu (iii) 1 mol Cu
  (iv) +1

32 d (i) 2880 coulombs (ii) 0.03 faraday e (i) 0.03 mol Na (ii) 1 faraday
  (iii) 0.015 mol Mg (iv) 2 faradays

Section 6 Thermochemistry and Enthalpy

34 c (ii) about $-4850\ kJ\ mol^{-1}$ e (i) 11 mol CH_4 (ii) 9790 kJ
  (iii) nearly 6 pigs

36 a 6400 J b (i) 0.01 mol C_2H_5OH (ii) $640\ kJ\ mol^{-1}$

37 c 24 d $38°C$ f (i) 24 g (ii) 80 kJ (iii) $38°$

38 b (i) $39\ dm^3$ (ii) $195\ dm^3$ (iii) 52 g c (i) 261 g d (i) $2.4\ dm^3$
  (ii) $1.2\ dm^3$ (iii) $2.4\ dm^3$

Section 7 Rates of Reaction

39 b 0.15 g c (i) 16 (ii) 40 e 0.5 M

40 d (i) 0.004 mol H_2 (ii) 0.262 g Zn 41 d $1/160$ mol O_2

42 c 2.5 minutes f (i) 0.1 g H_2 0.05 mol H_2 (ii) 0.05 mol Zn, 3.27 g Zn

43 d $3600\ cm^3$ O_2

Section 8 Reversible Reactions
46 b 21 g **c** (i) 96 dm³ (ii) 72 cm³
48 a (i) 91°C (ii) 190°C **b** 30 dm³ **c** 4 **d** (i) 4000 (ii) 1000 (iii) 59 kg
  e 2·95%
Section 9 Acids, Bases and Salts
50 a (ii) 0·25 **b** (ii) 2·65 g **c** (i) 20·4 cm³ and 20·6 cm³ (ii) 1/400
  (iii) 1/200 (iv) 20·0 cm³ (v) 0·25 (vi) 9·125 g
52 a (i) 1/40 (ii) 0·025 **b** 110
54 a 0·2 M **b** (ii) 0·005 (iii) 0·002 5 (iv) 0·355 g **c** (i) 0·125 M (iii) 0·60 g
56 b (iii) 880
Section 11 Metals
65 b 0·20 g **c** (i) 0·11 g (ii) 1·83 **e** 0·04 g
67 c (i) 12 dm³ (ii) 18·67 g (iii) 0·33
69 f (i) 3 faradays (ii) 108 x 10⁶ coulombs (iii) 1125 faradays
  (iv) 375 mol A1 (v) 10·125 kg
70 d 8·1 g
Section 12 Non-metals
74 h (i) 0·05 mol Fe (ii) 0·1 mol C1 (iii) $FeCl_2$
75 d (iii) 8 **79 d** (i) 48 g (ii) 64 g
82 b 4 g **e** (i) 200 g (ii) 20 g (iii) 20 g (iv) 4 g **f** 16 g
83 b 1·65 mol KNO_3/100 g H_2O, 80°C **c** 373 g NaC1 **d** 42·5 g $NaNO_3$
88 a (i) 85, 16·5% (ii) 80, 35% (iii) 60, 46·7% **b** 1050 g, 933 g
  c 19·0 kg, 21·3 kg
Section 13 Carbon Chemistry
92 f 34 250 kJ **93 d** (i) 12%
98 d 4 mol $C_{12}H_{22}O_{11}$ **e** (ii) 4 mol $C_{12}H_{22}O_{11}$ → 8 mol $C_6H_{12}O_6$ →
  16 mol CO_2 (iii) 384 dm³
Section 14 Materials and Energy Sources
102 a (i) 1·296 g (ii) 1000 cm³, 1.296 x 10⁻³ (iv) 14.4 **b** (ii) 10 cm³
103 a (ii) 1 : 1 **c** (i) 28 kg Fe_2O_3, 42 kg Ti_2O_3 (ii) 19·6 kg Fe, 28 kg Ti
  d 17·6 kg Fe, 23·8 kg Ti
105 a (ii) 4 dm³
107 a (iii) 5 cm³ **b** (iii) 2 mol e⁻ (iv) 230·4 kJ/mol H_2O
Section 15 Industrial Processes
109 a 100 dm³ **d** 1 : 2 or 1/2
Section 16 Miscellaneous
114 b (i) 19°C (ii) 6°C (iii) -7°C
115 c 230 000 tons (approximately) **116 c** (i) 90 g